STUDIES IN SOVIET HISTORY AND SOCIETY
General Editor: R. W. Davies

The series consists of works by members or associates of the interdisciplinary Centre for Russian and East European Studies of the University of Birmingham, England. Special interests of the Centre include Soviet economic and social history, contemporary Soviet economics and planning, science and technology, sociology and education.

Gregory D. Andrusz
HOUSING AND URBAN DEVELOPMENT IN THE USSR

John Barber
SOVIET HISTORIANS IN CRISIS, 1928–1932

Stephen Fortescue
THE COMMUNIST PARTY AND SOVIET SCIENCE

Philip Hanson
TRADE AND TECHNOLOGY IN SOVIET–WESTERN RELATIONS

Jonathan Haslam
SOVIET FOREIGN POLICY, 1930–33
THE SOVIET UNION AND THE STRUGGLE FOR COLLECTIVE SECURITY IN EUROPE, 1933–39

Peter Kneen
SOVIET SCIENTISTS AND THE STATE

Nicholas Lampert
THE TECHNICAL INTELLIGENTSIA AND THE SOVIET STATE
WHISTLE BLOWING IN THE SOVIET UNION

Robert Lewis
SCIENCE AND INDUSTRIALISATION IN THE USSR

Neil Malcolm
SOVIET POLITICAL SCIENTISTS AND AMERICAN POLITICS

David Mandel
THE PETROGRAD WORKERS AND THE FALL OF THE
OLD REGIME

THE PETROGRAD WORKERS AND THE SOVIET
SEIZURE OF POWER

Roger Skurski
SOVIET MARKETING AND ECONOMIC DEVELOPMENT

J. N. Westwood
SOVIET LOCOMOTIVE TECHNOLOGY DURING
INDUSTRIALISATION, 1928–52

Series Standing Order

If you would like to receive future titles in this series as they
are published, you can make use of our standing order
facility. To place a standing order please contact your
bookseller or, in case of difficulty, write to us at the address
below with your name and address and the name of the
series. Please state with which title you wish to begin your
standing order. (If you live outside the UK we may not have
the rights for your area, in which case we will forward your
order to the publisher concerned.)

Standing Order Service, Macmillan Distribution Ltd,
Houndmills, Basingstoke, Hampshire, RG21 2XS, England.

The Communist Party and Soviet Science

Stephen Fortescue
Research Fellow, Centre for Russian and East European Studies,
University of Birmingham

MACMILLAN PRESS in association with the
Centre for Russian and East European
Studies, University of Birmingham

First published 1986

Published by
THE MACMILLAN PRESS LTD
Houndmills, Basingstoke, Hampshire RG21 2XS
and London
Companies and representatives
throughout the world

Printed in Hong Kong

British Library Cataloguing in Publication Data
Fortescue, Stephen
The Communist Party and Soviet Science.—
(Studies in Soviet History and Society)
1. Science—Soviet Union—History—20th century
I.Title   II. University of Birmingham
Centre for Russian and East European
Studies   III. Series
509'.47   Q127.S65
ISBN 0–333–39419–4

0092701

335.430947

FOR

136226

To Alyona

Contents

List of Tables

viii

Preface

One of the most important features of the Soviet system is the Communist Party of the Soviet Union, not just as the single ruling party, but also as a ubiquitous apparatus of management and control. The intention in this book is to analyse the role of the Communist Party of the Soviet Union in the management of Soviet science. An attempt is made to determine the place the party gives science in the Soviet system and the general cultural and psychological conditions it creates for its conduct. But there is also a considerable amount of detailed study of the day-to-day management of science. No firm definitions of what science is are offered or employed, nor is any distinction made between science and technology. For the purposes of this study it could be said that anything that is done in a research institute is science or part of the scientific process. The empirical work is based on civilian research, and the conclusions probably are not applicable to the defence sphere. Both the natural and social sciences feature in the analysis. I have generally drawn out the differences in their treatment in the Soviet Union, although not in a rigorous way. The time focus is also vague. I have tried to be aware of the historical background, but the study is essentially concerned with the contemporary situation.

I have attempted in the Introduction and Conclusion to put the work in a theoretical framework, in terms of the current debate about which model can best be used to understand and analyse the Soviet system. However I hope that the empirical analysis is valuable even without the theoretical trappings.

The book was researched and written while I was on a Postdoctoral Fellowship in the Department of Political Science, Research School of Social Sciences, Australian National University. I am grateful to have had the opportunity to work in such relaxing and simultaneously stimulating conditions, and am particularly grateful to the Department and School for a few months' breathing space when unemployment was threatening. The threat was removed by the Centre for Russian

ix

and East European Studies at the University of Birmingham, in which
excellent institution the book has been completed. My Fellowship in
the Centre is funded by the Economic and Social Research Council.
  My deepest thanks go to T. H. Rigby for friendship, assistance and
advice going right back to the beginning of my interest in Soviet
studies. R. F. Miller has also been a constant friend and colleague,
and has taken a particular interest in this work. Peter Kneen has
provided detailed comments and suggestions, as have Archie Brown,
John Miller and Alec Shtromas. Greg Topchian and Russell McCaskie
have done much of the hackwork.

Birmingham   STEPHEN FORTESCUE

Introduction: A Theoretical Framework

It probably would have been easier and safer to make this a purely empirical study, a straightforward description of the formal and informal behaviour of the party institutions and personnel involved in science management. However, two factors have encouraged me to devote some attention to theoretical questions. Firstly, there is the degree of disagreement in the West on the correct theoretical approach to the study of Soviet politics. This has reached the stage where one feels almost obliged to take a stand. Certainly, the issues raised must be considered by anyone claiming to be interested in gaining a genuine understanding of the way the Soviet system works. Secondly, two almost contradictory approaches can be taken to the study of Soviet science. We can concentrate on the ideological and political control of science, something which will produce shocking results for those reared in the Western tradition of 'free' science. Or we can concentrate on the involvement of powerful science bodies like the Academy of Sciences in the struggle for bureaucratic power and influence. This could well produce a picture that will look more familiar to those who know Western science politics. Which approach we take will very likely be determined by a preconceived view of the way the Soviet Union works, and the findings will almost certainly confirm that preconceived view. For a more balanced approach, we must be aware of the links between theoretical models and empirical approaches. We might still arrive at the same conclusions, or we might find ourselves unable to come to a firm, non-contradictory conclusion. But at least we will know that we have considered all the possibilities.

No brand new theoretical model will arise from this study. Nor will the empirical work be done in terms of a single model chosen in this Introduction. Rather, the empirical research will be presented in straightforward descriptive terms, and then in the Conclusion an examination will be made of which model is best fitted to the empirical results.

I have chosen to divide models of the Soviet system into three categories: totalitarian, 'vanguard party' and pluralist.

TOTALITARIANISM

The totalitarian model in its most formal guise has long since lost its dominance. The lists of essential features drawn up, for example, by Friedrich and Brzezinski, have had to be so qualified and modified to suit changing Soviet reality that they have been largely abandoned,[1] while competing models which deny even the basic concepts of totalitarianism have gained enormous popularity. However, some leading Sovietologists, such as the late Leonard Schapiro in Britain and Boris Meissner in Germany, never discarded the model,[2] while in recent times there has been a revival of interest in totalitarian concepts, linked to some extent to the influence in the discipline of Soviet emigres. Their concerns are not the traditional ones of the specific details of social atomisation and the mechanics of the police state, but rather with a situation in which, whether for the protection of the personal interests of highly placed individuals or for the implementation of a degenerate ideology, all social activity, scientific research included, is totally controlled, indeed repressed at its very roots, by the party leadership. Some specifically claim that it is not the abuse of power that produces the inhumanity of the totalitarian system, but rather the relentless projection of an impoverished image of humanity, through an ideology from which all ethical considerations have been removed, and mechanisms of bureaucratic control that emphasise the powerlessness of the common man.[3]

Thus while they base themselves very much on personal experiences, the main concern of those employing the neo-totalitarian approach is with the moral and spiritual atmosphere of Soviet life, rather than with institutions and political processes. The approach is not presented as a self-conscious political theory, but its essential features can be gained from a reading of the enormous emigre literature of recent years.

When applied to science, the emphasis of traditional totalitarian theory tended to be on ideology, with many Western observers being convinced by the ideological packaging of the activities of such people as Lysenko that ideology lay at the bottom of the destruction of Soviet genetics and the attacks on other scientific disciplines.[4]

But over the last twenty years, with a much-needed re-evaluation of the role and motivations of people like Lysenko and the obvious withdrawal of the party from such interference in the natural sciences, ideology is no longer the most common totalitarian explanation of the driving force behind the party's control of science.[5] Many analysts now see Marxist–Leninist ideology as little more than a language of political behaviour, adhered to out of habit, fear or cynicism. This is well summed up by Aleksandr Shtromas:

> People were no longer asked to believe in Communism or to 'internalize' its values but simply to behave according to the appropriate rules and customs of a Communist regime. It was no longer necessary for people to accept Communism as an ideal so long as they accepted it as the socio-political framework of reality within which to further their personal interests and goals. The pursuit of these personal interests and goals was thus tacitly, if partially, recognised by the authorities as a valid basis for socially approved behaviour.[6]

This new concept of the role of ideology has given rise to the 'bureaucratic corruption' view of totalitarianism. Particularly among emigre writers with a lifetime's experience of the Soviet system, the usual claim is that the party is destroying Soviet science through its cynical incompetence, backed up by the corrupt surrender to selfish careerism of the leading members of the scientific establishment. This is best summed up in the English title of Mark Popovsky's book, *Science in Chains*.[7] A number of emigres from the Soviet scientific community have provided considerable empirical data on the methods of 'bureaucratic corruption', and this gives us the basis for an examination of the neo-totalitarian model in terms of institutional and political processes, rather than purely in terms of moral and spiritual conditions.[8]

There are those professing the totalitarian view who are not prepared to admit any successes for Soviet science whatsoever – apparent successes are either the result of espionage or the economically irrational overconcentration of resources in one field.[9] If any individual scientist seems to have won for himself independent status and prestige, that is only because he has sold out to the party and should therefore be considered part of the bureaucratic elite rather than a member of the scientific community.[10]

There are others who are prepared to see the totalitarian model in less categorical terms and as being compatible with other models. Aleksandr Shtromas, for example, is attracted to interest group theory as a useful working model, but strongly insists that this in no way denies the totalitarian essence of the Soviet system, with all the consequences of the cynical and debilitating adherence to an obsolescent ideology, the oppressive presence of a hidebound party apparatus in all areas of activity, and a strong leavening of irredeemably orthodox 'partocrats' in all Soviet institutions.[11]

To determine whether party control of Soviet science is of a totalitarian nature, we have to answer two questions. Firstly, do the strictures of Marxist–Leninist ideology still significantly affect the type of research that is done in the Soviet Union? Secondly, are the values of 'bureaucratic corruption' so dominant or so successfully imposed on Soviet science that there is no room for free research, creativity or originality within that science and no possibility of the scientific community providing an independent input into the policy-making process?

THE 'VANGUARD PARTY' MODELS

Vanguard party models are conceptually very different from the totalitarian model, although to some extent the same empirical findings could be fitted quite well to either. The totalitarian model sees the existence of groups in society as either impossible (in the most extreme social atomisation versions) or irrelevant (in 'partocrat'-type versions, in which totalitarian values are held and power exercised by a broad range of individuals throughout society regardless of functional or institutional affiliation).

In vanguard party models far more attention is paid to functional and institutional groups. Those putting forward such models feel that the empirical importance of such groups cannot be denied. They can be clearly identified in terms of institutional and legal status or the common backgrounds and functions of individuals. It is then claimed, whether deductively or on the basis of empirical evidence, that such groups have their own interests and that to a large extent the behaviour of the groups can be predicted on the basis of those interests.

But the essential feature of vanguard party models is that these groups do not wield any genuinely autonomous power in the political system. They are totally dominated by the party, usually meaning the 'party leadership' with the party apparatus as its operational arm.

The 'party leadership' can be variously described as the General Secretary, the Politburo, the Central Committee, or some dominant faction in those bodies. The party has such resources at its disposal that its control of all other groups is in the final analysis total. They either exist purely to serve the party, or they are in no position to resist it when their own interests conflict with those of the party.

The resources at the disposal of the party are numerous, if granted different weight in different versions of the model. The party's control of the only acceptable ideology in the Soviet system is not seen generally as having enormous practical significance. Its control of the forces of repression and coercion, although like ideology not ignored, is usually recognised as being an insufficient explanation on its own of the party's dominant role. Indeed, the stress has shifted to seeing the party as possessing considerable authority, in the sense of power based on legitimacy.[12]

This authority is, however, backed up by such bureaucratic powers and tactics as control of personnel through the *nomenklatura*, cooptation of 'outsiders' into the party elite,[13] strict control of the setting up and granting of powers to formal institutions, dominance of policy-making structures and procedures, and of course a ubiquitous apparatus and membership. Some of the best-known Western vanguard party models combine this strong belief in the dominance of the party with what they see as the essentially bureaucratic nature of Soviet politics, to produce a description of the Soviet political system as one big bureaucratic organisation. One such model is Alfred Meyer's 'USSR Incorporated'. Meyer likened the Soviet political system to a Western business corporation, with the Politburo or party leadership equivalent to the Board of Directors. All the line and staff agencies of the corporation, although they might have their own interests and sometimes disruptively assert them, exist essentially to serve the goals of the corporation as set out by the management, usually in terms of the 'corporate culture', Marxism–Leninism.[14] Such a model provides enormous scope for political activity of all types, including group politics, but allows us no doubt that such politics is dominated and determined by the party.

A similar, although less metaphorical, approach is taken by T. H. Rigby, with his idea of the 'mono-organizational society'. Rigby has described the Soviet system in the following terms:

> The most salient feature of communist systems is the attempt to run the whole society as a single *organization*, in which almost no socially significant activities are left to autonomously interacting

individuals and groups, but instead are managed by centralized, hierarchical agencies, themselves subject to close coordination, principally by the apparatus of the party . . . . Societies with these characteristics might be called *mono-organizational*.[15]

From this quotation it is clear that Rigby grants a role to institutional groups, his 'centralized hierarchical agencies', but that they are not free actors. He writes in another place:

The political leadership in such a system is not content to 'rule', 'govern', or 'control', leaving much of the day-to-day activity of society to be determined by autonomous or quasi-autonomous subsystems of traditional, associational, or market type. It sets out to run the society as a whole much as a business corporation is run by its top management or an army is run by its commander-in-chief and general staff.[16]

From his empirical writings, it is clear that Rigby, like Meyer, by no means intends to deny the existence or liveliness of political competition and conflict in the Soviet Union.

The most recent corporatist theories can also be included in the vanguard party category, with their stress on the role of the central government in setting up a system of cooperating groups. Phillipe Schmitter's definition of corporatism sounds not unlike Rigby's 'mono-organizational society':

Corporatism can be defined as a system of interest representation in which the constituent units are organized into a limited number of singular, compulsory and non-competitive, hierarchically ordered and functionally differentiated categories recognized and licensed (if not created) by the state and granted a deliberate monopoly within their respective categories in exchange for observing certain controls on their selection of leaders and articulation of demands and supports.[17]

Nevertheless, there is room for bargaining between the political leadership and 'licensed' groups. Again the process is dominated and manipulated, if not created, by the political leadership. Schmitter's ideas have been applied to the Soviet Union, particularly by Valerie Bunce and John Echols.[18]

One of the most interesting variants of the vanguard party model is the official Soviet version of the 'leading role of the party'. The 'leading role of the party' is the basic principle of Leninist politics, and is recognized as such by the Soviet Constitution and the Party Rules.[19] The party, being officially considered the vanguard of originally the proletariat and now the whole people, has a special claim to preeminence in decision-making. However that preeminence is within the framework of the recognized legitimacy of other sectional groups. There is a quite well-developed Soviet interest theory which admits the existence of the special, competing and even selfish interests of these sectional groups.[20] With the increasing inappropriateness of 'administrative methods' the essence of the political process is described as the manipulation of these interests in such a way that the social interest is advanced. B. M. Lazarev expresses the essence of the matter most bluntly:

> The state, struggling with narrow sectionalism (*vedomstvennost'*), often uses sectional interests, to the benefit of the social, 'guiding' one sectional interest against another which is adversely affecting the resolution of some social problem.[21]

Here one should read for the state the party.[22] A similar idea has been expressed more recently by a far more authoritative figure. Andropov declared in *Kommunist* in 1983:

> The interests of society as a whole are a most important reference point for the development of an economy based on socialist property. But this does not mean that in the name of the common good socialism suppresses or ignores personal and local interests and the specific demands of different social groups . . . . One of the most important tasks in the perfecting of our economic mechanism is to develop a precise accounting of these interests, to achieve their optimal blending with the general interests of the people, and in such a way use them as the motive force in the growth of the Soviet economy, raising its effectiveness and the productivity of labour, and an all-round strengthening of the economic and defence might of the Soviet state.[23]

The party is able to attain this 'blending' of interests through its right to give guiding directions (*rukovodiashchie ukazaniia*) on all major issues, its personnel powers and powers to verify fulfilment of its decisions, as well as its right to interpret and apply the tenets of Marxism–Leninism.

The 'leading role of the party' provides a useful framework even for Western scholars who accept the dominance of the party in the Soviet system.[24] However few would accept the implication in Soviet definitions of the party's 'leading role' that the party plays a progressive role in society. In fact 'the leading role of the party' is seen to be used more often than not in an entirely gratuitous way as a cover for the straightforward struggle for and maintenance of personal power and privilege, entailing the violent suppression of all other social forces.

It does seem possible at least to hypothesise, however, that the party leadership has some commitment to 'social progress', even if that means no more than strengthening national defence, improving the working of the economy, and perhaps, as a residual, improving the living standards of the population, and, even if the only motivation for such a commitment is that it is necessary to be seen to be 'getting the job done' in order to maintain one's position in power. At present scientific and technical progress is seen as the major determinant of 'social advance', and the apparently genuine commitment of the party leadership to science and technology is therefore part of the progressiveness of its 'leading role'. Further, because sectional interests exist that are less committed to science and technology, the party involves itself in a political process characterised by competing interests in a way already described by B. M. Lazarev, ' "guiding" one sectional interest against another which is adversely affecting the resolution of some social problem', in this case technological progress.

If we are to accept the vanguard party model we would expect to find a political process of competition between groups with different sectional interests. This competition would, however, be overseen, manipulated and eventually brought to a satisfactory conclusion by the party. We would therefore expect to see the party leadership making the major policy decisions, presumably in a reasonably forceful way, and its apparatus working solidly to ensure the fulfilment of these decisions. Certainly, in the long term one would expect to see the party successfully implementing, for better or for worse, most of its own policies.

Nevertheless those adhering to vanguard party models do not expect to see the party always victorious in its efforts to manipulate the political process. The Soviet authorities regularly admit to their failures, or as Rigby puts it:

> This attempt [to run all social activies from a single centre] is only partially successful . . . . In an organization that seeks to embrace a whole complex society of continental proportions, the incongruencies between its authority and action levels . . are likely to be all

the more marked . . . . Nevertheless, the fact that the authority level is patterned as a set of command-hierarchies *running* almost all activities under a single ultimate direction appears to be the best starting point for attempting to comprehend what goes on in Soviet-type systems.[25]

However, such a model makes little sense unless we can expect the party to get its way in the great majority of cases, and to get its way because of its special place in society. As long as all other groups in society are firmly under its control this is a reasonable expectation.

PLURALIST MODELS

Pluralist models pay more critical attention to the status of sectional groups. Those developing such models claim, usually on the basis of the increasing complexity and specialisation of modern society, that sectional groups arise in answer to social, economic or political needs that cannot be denied, ignored or even manipulated by the political authorities. The earliest precursor of such a view, in the Soviet field, was Barrington Moore, with his speculation in 1954 on the rise of the technocracy:

> Since the Soviet dictatorship cannot dispense with the rational element, the interesting question arises of whether or not the dictatorship might be rationalizing itself, so to speak, against its will. Perhaps the Soviet system is suffering from the disease of creeping rationality, much as some people think ours is suffering from creeping socialism. Beneath the turmoil of the purges, demotions, and meteoric promotions that constitute such a prominent feature in the Soviet political landscape one may detect throughout the history of the regime a steady lumbering movement in the direction of the clearer allocation of responsibility. It is worthwhile, therefore, to explore what might happen if this trend should display still greater strength with the passage of time.
>
> The end result can be imagined as technocracy – the rule of the technically competent.[26]

The concept of technocracy – the rule, whether it be *de facto* or *de jure*, of those with specialised knowledge – has a very long history, going back as far as Plato.[27] Since Barrington Moore's original suggestion in 1954, the basic concept of technocracy has taken a strong hold on Sovietology. Technocratic categories lie at the base of many studies in the broad fields of Soviet political leadership and economic management. Indeed they have probably gained wider use in the study of the Soviet Union than of Western countries. Whereas in the West politics as a profession has generally maintained a unique and independent character, with technical knowledge rarely being considered a necessary prerequisite, in the Soviet Union, particularly since the elimination of the Old Bolsheviks, there has been no specific category of people with their own peculiar background, education or culture who could be called professional politicians. The nature of the Soviet system, as it has developed at least since the 1930s, means that there cannot be the same functional and cultural differences between politicians and administrators as there are between politicians, civil servants, private sector managers and technical specialists in the West. This makes it far easier to see modern Soviet government in terms of rule by 'bureaucrats', 'managers' or 'specialists'.[28]

For our study it is worth considering two major points usually made by technocracy theorists. Firstly, that in virtually all societies, and perhaps particularly those that claim to have faith in continuing human progress, high status usually obtains to scientists and specialists in general. A number of observers have pointed to a long-standing commitment to science and technological progress among the Bolsheviks, and indeed see that commitment as constituting a major element of legitimacy for the Soviet leadership.[29] The logical implication is that the leadership is particularly likely to grant status and influence to scientists in science policy-making and even in other areas of policy-making.

There is another strand of technocracy theory which claims that it does not matter whether the political leadership has a commitment to science or not. Because economic development, national defence and, to a great extent, social control depend on scientific and technological advance, specialists are able to make enormous demands of political leaders and in effect blackmail them into granting these demands. Firstly, the politicians do not have the technical knowledge to argue with specialists who claim their latest work to be vital; secondly, there is an inexorability in the logic of scientific advance – particularly in a competitive world – that is impossible to resist ('if we don't do it someone else will').[30]

These are questions that could be applied to the Soviet system most profitably, particularly in a study of the relationship between the party and science. However, most students of Soviet politics would consider that there is a critical lack of differentiation in the technocracy thesis, that technocracy is too broad a concept to apply to all the myriad institutions that are involved in scientific and technological development, while it entirely ignores what might be important groups in society that have no connection with such issues.

Nevertheless, the idea that in a complex society there is room for groups with such strategic positions that the political leadership is unwilling or unable to ignore them forms the basis of pluralist models. Unlike the technocracy theorists, however, those using these models see a considerable number of groups representing a wide range of interests and values occupying strategic positions. Some of the more extreme variants of pluralism have gone so far as to deny an independent role to the party in the direction of Soviet society. Thus M. P. Gehlen writes that the party elite has become 'a composite of elite representatives of various groups rather than a separate group acting as a supreme arbiter over all inferior social units'.[31] Hill, Dunmore and Dawisha also stress the carefully balanced representation of all important sections of society in leading organs.[32] According to this view, the party as a political actor is no more than the party apparatus, with its own special sectional interests, represented in leading policy-making organs just the same as other sectional groups. It pursues its interests on essentially an equal footing with other hierarchies and groups, the outcome being determined by the relative political strength of the various contending groups. (Many would deny, of course, that even the party apparatus makes up a single homogenous group.)[33]

Most pluralist theorists see such a view as too extreme. They grant the party a special status in the political system, and see leading organs such as the Politburo as specifically party bodies, not simply the forums for group interaction. Thus the party leadership represents political authority, much as the ruling party and its Cabinet represent political authority in a parliamentary democracy (although almost all would claim that in the Soviet case the power is far greater and more extensive).

However, such pluralists do not see the party as having the same dominant position as vanguard party theorists do. Sectional groups cannot be simply manipulated by the party into serving its own interests. In policy-making the party will often rely on sectional groups to provide the initiative and essential input, and indeed could find itself having to accept input it might not have wanted. Then, at the policy implementation stage the party will not necessarily find itself able to enforce policies that have been decided. Pluralist theorists would consider that these phenomena play such a frequent and significant role in Soviet politics that it becomes inaccurate to grant the party the dominance that the vanguard party models imply. They consider that there are fundamental features of a modern political system that make such a degree of dominance virtually impossible.

A considerable amount of interesting case-study work has been done with something approaching a pluralist approach (although many of the authors would reject some of the conclusions I am suggesting here). These include the pioneering article by Schwartz and Keech on the 1958 educational reform,[34] Thane Gustafson on environmental issues,[35] and Peter Solomon on the making of criminal policy.[36] John Lowenhardt concentrates his attention on the 1961 Academy of Sciences reorganisation, but also collects a large number of previously published case studies and subjects them to theoretical analysis.[37] In recent times, without the full theoretical conclusions that I have just sketched being drawn, the approach has even been used to analyse the policy-making process under Stalin.[38]

Earlier pluralist writers tended to base the groups which made up the pluralist system on a variety of factional, patronage, functional and opinion grouping definitions.[39] There was some unwillingness to admit that the party authorities would allow sectional interests to organise institutionally. This derived from a determination to avoid accusations of a naive acceptance that public protest groups were allowed to function in the Soviet Union.[40] However, in recent years the consensus seems to be that Soviet sectional groups are in fact usually based on the society's major institutions, as suggested by Jerry Hough's nomenclature 'institutional pluralism'.[41]

There are a number of implications of 'institutional pluralism'. Firstly, if we equate interest group with institution, we therefore equate interest with institutional goal, or as it is usually known in the West, 'organisational goal'. Organisational goals are a somewhat controversial issue in the Western literature;[42] in a system apparently so dominated by a single ideology it is bound to be even more controversial and, one might expect, rather difficult to identify. The second implication follows from the first to some extent. If the group's activities are determined by its organisational interest or goal, the process of how that interest or goal is defined becomes crucial. As long as the group has some independence from the political authorities, that process will be essentially an internal political process. It would seem that the pluralist approach puts the emphasis squarely on the possibility of conflict and political behaviour within institutions, not, as some have claimed in criticising the approach, on assumptions of internal unity and homogeneity.[43]

The third implication is that groups, or potential groups, that do not have an institutional existence could be expected to do poorly in the political process. This becomes relevant in Soviet science, given, for example, the relative weakness of the top-level institutional representation of the branch science sector and the special history of the development of various social science disciplines.[44] Fourthly, we have to ask ourselves an important theoretical question. The pluralist approach grants institutional groups a degree of operational independence from the party authorities. One would like to extend the point to the setting up of the institutions. However, it would be very difficult to claim that Soviet institutions set themselves up without party approval or initiative. Does the apparent control by the party of institutionalisation and its use of that control to structure the political process destroy the validity of the pluralist approach? Can we save the situation somewhat by relying on the basic argument of technocratic determinism, that the party has no choice but to grant essential specialist groups an institutional existence, whether it is demanded explicitly or implicitly? The final implication of the institutional approach to pluralism is that we would expect factional, patronage and non-institutional functional or professional groups, as well as individuals, to play a reduced role. As we will see later, the last two – professional groups and individuals – are of particular relevance to science.

If we are to accept the pluralist model, in the science field we would expect to find scientists directly involved in the policy-making process, probably on an institutionalised basis and with an initiating role. We would also expect to find evidence of the inability or unwillingness of the party leadership to take on major non-party institutional groups and its inability to enforce the implementation of policies that are arrived at. The latter implies serious operational weaknesses of the party apparatus.

Thus we have three groups of models: the totalitarian, which claims that the party exerts total dominance over the Soviet political system by denying the legitimate existence of any values other than its own repressive ones; the vanguard party, which claims that the party asserts its dominance through the more sophisticated means of controlling and manipulating a whole range of legitimately existing sectional groups; and the pluralist, which denies the party any inherent dominance, due to the inevitable power held by sectional groups in complex societies.

These groups of models are perhaps not entirely mutually exclusive. Nevertheless, it should be possible to choose one at the expense of the others. It will be one of the aims of this study to do so.

1 The Ideology of Science

This chapter is designed primarily to set the scene for the 'hard' empirical work to come in later chapters. No claims are made as to the originality of the propositions contained in it, and indeed I have relied to a large extent on Western secondary sources. Nevertheless, an attempt is made to answer four important questions that are of direct relevance to our consideration of the totalitarian model and the technocratic aspect of the pluralist model. Those questions are: firstly, is Soviet science severely restricted by the demands of Marxist-Leninist ideology; secondly, and to some extent alternatively, is there a Bolshevik 'ethos of science' that gives science a special status in the eyes of the political leadership; thirdly, regardless of the answer to the first two questions, is science ground down by an oppressive atmosphere of political and bureaucratic control; fourthly, do scientists collectively have specific values which they hold in spite of and as a form of defence against more oppressive political values? If we can answer either the first or third question in the affirmative, a totalitarian conclusion could well be justified. If, however, they are answered in the negative and we find some truth in the propositions contained in the second and fourth questions, we have at least laid the basis for a pluralist conclusion.

MARXIST–LENINIST IDEOLOGY AND SCIENCE

Firstly, let us look at the formal status of science within Soviet Marxism. The natural sciences have never been the subject of controversy in Marxism to the degree that such matters as surplus value, class structure or the stages of social development have been. But until Stalin laid all such questions to rest in the early 1930s, the status of science was nevertheless a matter of virtually constant debate among Marxist philosophers and theoreticians. The essence of the debate was the contradiction between metaphysics and positivism in the philosophy of science of Marx and Engels – whether science (including Marxism itself) is an objective reality determined by the laws of history independently of the intervention of mankind, or whether it is inevitably modified or coloured by those individuals who practise scientific research and theorising. The latter view would suggest that theoreticians, particularly philosophers, have a role to play in interpreting objective reality and the distortions of it by humans. Although Marx and particularly Engels, and then Lenin after them, tended towards the metaphysical view – that science as we see it is coloured by those practising it and therefore is a legitimate concern of philosophers – they left enough ambiguity in their positions that the debate continued in the Soviet Union into the 1920s.[1]

During the 1920s the question was hotly debated by two factions. The 'mechanists' took the view that the natural sciences are inherently and naturally dialectical materialist, regardless of the ideological world view of the scientists involved, and that they are therefore not subject to the control or interpretation of philosophers. Indeed, the social sciences, if they are to become truly scientific, must model themselves on the natural sciences and will eventually merge into them. The mechanists, who were largely concentrated in the fringe area between philosophy and natural science and were concerned to give ideological respectability to the 'bourgeois' scientists who dominated Soviet science at the time, were opposed by the philosophers of the Deborinist school. Not surprisingly, considering that they were philosophers, the Deborinists maintained that there was nothing inherently dialectical materialist about the natural sciences, particularly if they were being studied by 'bourgeois' scientists with all sorts of idealistic and religious ideologies. Therefore, philosophers had to determine whether particular scientific theories were in fact dialectical and materialist and therefore contributing to social and economic progress.

The debate was seen, particularly by the Deborinists, as being essentially a philosophical one, and they showed no particular desire to apply their theories in practice. Thus, while the party authorities showed vague signs throughout of tending towards support for the Deborinists, the debate was largely ignored by scientists themselves and seems to have had no effect on any research in the natural sciences. This seems to be true even after the Deborinists had 'won' the debate, with mechanism being declared revisionist in 1929. (It can be inserted here that the 1920s were also a period of relative freedom in the social sciences, or at least of open debate between competing points of view.)

The Deborinists did not enjoy their victory for long. The party authorities, particularly the now dominant Stalin, did not like the suggestion in the Deborinist case that philosophy, and professional philosophers, were the supreme arbiters of Marxist ideology and therefore Soviet reality. In 1931 the Deborinists themselves were declared to be revisionist. This did not mean that the mechanists were able to make a comeback. Science was still subject to philosophy, but now philosophy itself was subject to the changing political needs of the party leadership.[2] There was nothing particularly new in that, but the crudity of the party's new manipulations of philosophy and the ruthlessness with which any hint of dissent was suppressed meant that virtually all intellectual content was removed from Soviet Marxism.

The ideological implications for the natural sciences were clear – not only was science subordinate to philosophy, but through philosophy it was totally subordinate to the party leadership. Only that science was genuine science that had the approval of the party; science that was condemned by the party was not real science. The immediate practical implications are less clear cut. The vicious anti-bourgeois specialist campaign and the *vydvyzhenie* movement (a programme vastly to increase the numbers of proletarians and peasants moving into science and to exclude people of other social backgrounds), which had been in existence since 1928, were largely ended in 1931, leaving the 'bourgeois' specialists still in command of Soviet science; despite the efforts of a few eager and ambitious young scientists to 'Bolshevise' the natural sciences as philosophy had been 'Bolshevised', the party drew back; and while Stalin and his spokesmen lambasted capitalist science as necessarily false and inferior, he continued to industrialise largely with imported technology.

The 1930s did see, of course, the rise of Lysenko, one of the grimmest chapters in the history of Soviet science.[3] By the late 1940s he was completely dominant in Soviet biology, while his methods were being increasingly used in other fields.[4] Lysenkoism was a classic case of party approval determining which science was real science. However, it is now generally accepted that Marxist–Leninist ideology had little to do with Lysenko's dominance. Although Lysenko and his supporters, particularly his chief ideologist Prezent, were concerned to demonstrate the ideological superiority of his version of biology, the major reasons for his success were his skilful manipulation of science politics, his charismatic qualities and his extravagant promises of enormous practical benefits.[5] Throughout the Stalinist period it was perfectly clear to all that party approval, when the party chose to involve itself, was virtually the sole arbiter of what was genuine and permitted science and what was not. While it was considered important to protect oneself as much as possible against ideological attack from philosophers or professional competitors, it was also recognized that party approval did not in fact depend ultimately on ideological factors, but rather on the ability of scientists to play politics and demonstrate practical benefits.

There has been no fundamental change in the formal ideological situation since the death of Stalin. While the strident language of the Stalin era is no longer in evidence, and particularly since the fall of Khrushchev and the final disgrace of Lysenko the authorities have been concerned to reassure the scientists of their immunity from such 'excesses', nevertheless there has been no reversal of the twin positions that socialist science and those that practice it must be subject to the philosophical principles of Marxism–Leninism, and that the sole guardian and true interpreter of Marxist–Leninist philosophy is the Communist Party. However, there have been changes in the formal ideological status of science in the last thirty years which, while not removing this fundamental principle, have to a limited degree given science greater independence from formal philosophical and ideological control.

Firstly, there was the declaration in the early 1960s that science 'is becoming a direct productive force'. Marx and Engels did not devote much attention to the place of science in the social structure they devised. While Marx seemed to assign science to the base rather than the superstructure,[6] Engels apparently allocated it to the superstructure.[7] Russian Marxists, faced by slow scientific development in their homeland, with a generally voluntaristic approach to social and economic development, and relying on a scientific community dominated by 'bourgeois' specialists, tended to see science in superstructural terms, meaning that its nature was determined by class relationships and could be manipulated by the ruling class of the day, including the Bolshevik-led Soviet working class after 1917. This tendency was strengthened and made more explicit by the Deborinists and then by Stalin and his 'Bolshevising' philosophers. They placed enormous emphasis on the inherent differences between 'bourgeois' and 'socialist' science and the complete subordination of the latter to the will of the Bolshevik party.

Oddly enough it was Stalin himself who first raised the possibility of changing the status of science, in his famous 1951 work on linguistics, *Marxism and Linguistics*.[8] He noted that language is connected with both man's productive activities and other social and intellectual activities that belong to the superstructure. Thus language, and the science of language, linguistics, belong to neither the base nor the superstructure. This opened the way for a debate among Soviet scholars, mostly philosophers, on the place of science in general, including the natural sciences. A number of the participants in the debate were prepared to allocate the natural sciences to the base, some separately from the social sciences (F. V. Konstantinov), others together with the social sciences (S. Strumilin).

These views were given a great boost by the increased attention given scientific and technological development by the post-Stalin leadership. This change is generally dated from Bulganin's speech at the July 1955 Central Committee plenum.[9] The ideological implications of this new emphasis on the role of science in developing the productive forces of socialist society and the party's resulting commitment to bring about a massive improvement in Soviet research and its application to production were given official status in the 1961 Party Programme, in which it was stated:

> The Party will do everything to enhance the role of science in building communist society. It will encourage research and discover new possibilities for the development of the productive forces, and the rapid and extensive application of the latest scientific and technical achievements . . . Science will itself in full measure become a direct productive force.[10]

This stage will be reached when science and production become totally assimilated with each other. On the one hand, science will no longer be something which only responds, after a lag, to the demands of the production process; instead it will work together with production elements to ensure a constant improvement in production methods. On the other hand, there will be no science, no matter how theoretical, which will not have a direct connection with production. This will arise only in a true socialist society, in which there is no scientific 'false consciousness' producing metaphysical and idealistic research of no practical benefit, and no contradictions between the interests of physical labour in production and the intellectual labour of scientists.[11]

The benefits to the scientists of this declaration were twofold. Firstly, science, although only able to fully realise its potential under socialism, does have continuity going right back through human history. To the extent that it is a productive force rather than part of the superstructure – that is, to the extent that it contributes to the development of the forces of production – any science, no matter in what society it originates, is potentially useful. This officially sanctioned the already greatly increased involvement of Soviet scientists in the international scientific community. Secondly, because science was still only 'becoming' a productive force, it required massive expenditure and personnel increases, indeed at a rate outstripping industrial investment, in order to catch up to the level already gained by the traditional forces of production. The scientists wasted no time in pushing such claims.[12]

For that reason the scientists might not have been altogether happy when in 1966, at the 23rd Party Congress, Brezhnev declared that 'the transformation of science into a direct productive force has fundamentally changed the character of man's work'.[13]

With the arrival of developed socialism, the stage was set for the complete removal of the contradictions between physical and intellectual labour, the link between science and production was total, and Soviet science had developed to such a level that it could operate as a fully fledged member of the international scientific community. But this achievement came at the same time as a reduction in the funding priorities given to science. With science having become a direct productive force, it was now possible to adopt a strategy of 'intensive' rather than 'extensive' growth, that is, making better use of existing resources rather than allocating ever greater new resources. A slowdown in resource allocations to Soviet science was indeed seen at this time (made necessary by enormous demands for investment in other areas at a time of economic slowdown).[14]

The second area of change in the ideological place of science in Soviet society, and one perhaps more interesting than arid Marxist discussions of base and superstructure, has been the extensive and varied debate in the last twenty years on the so-called 'scientific and technical revolution' (STR).[15] With the granting to science of potential status as a direct productive force being a prerequisite for serious discussion of the STR, it came into its own as a subject of academic debate and analysis in the 1960s.[16] The original discussion focused on the role of scientific revolutions in an historical context (centred in the Institute of the History of the Natural Sciences and Technology, IIET),[17] and the technical aspects of changes in economic and industrial management that would come from scientific and technological developments (led by the Ukrainian cyberneticist G. M. Dobrov).[18] Both approaches placed science very much at the centre of economic progress, and forsaw exciting prospects for the development of human organisation in the future. However, neither put their analyses in a specifically political context, and indeed showed a 'technocratic' bias (a tendency to neglect the role of the party) that meant the concept had to undergo a transformation before it could be adopted as the basis for a political strategy. In the early 1970s Brezhnev adopted the STR as a slogan to be used to shake the established industrial bureaucracies out of their anti-innovatory ways. New theorists of the STR, with impressive political connections and skills (people such as D. M. Gvishiani and V. G. Afanas'ev), came on the scene and presented the STR in more politically acceptable terms.[19] This included claims that technical development cannot be regarded as an end in itself, but must be related to social development;[20] that technical development can bring about desirable social development only if it is directed and controlled by a progressive political force, that is, the CPSU; and that the STR, while able to realise its true potential only in socialist conditions, is nevertheless a universal phenomenon (this was an important part of the economic side of Brezhnev's *detente* policy in that it ideologically justified massive technology imports from the West). Brezhnev's patronage of the concept led to an enormous boom in coverage in the Soviet press, with the STR promising changes in Soviet society ranging from the mundane to the fantastic.[21]

The STR is mentioned more rarely today, and has been largely replaced by the less emotive phrase 'scientific and technical progress'. This is presumably because the practical results of the STR have been less than revolutionary and the deterministic and technocratic aspects of the concept have continued to worry the leadership. Nevertheless the STR has by no means been abandoned,[22] and its main effects are lasting. It has provided a framework for the serious analysis of the place of science, including Western science, in social and economic development outside the arid and rigid Marxist framework of base and superstructure and the Stalinist categories of socialist science and capitalist pseudoscience. That analysis has determined that science is a universal phenomenon, even if one which can only realise its potential in a socialist society; it provides limitless scope for development right up to the achievement of communism; it gives a place to social factors in economic and technological development; and while the success of the STR is inevitable it is something which has to be worked hard for by a committed party and government.[23]

These changes in the ideological status of science should be seen in the context of an apparently genuine determination by the authorities not to have any repeat of Lysenkoist-type phenomena in Soviet science. Lysenko was, it seems, the second victim of the October 1964 coup, with, according to Zhores Medvedev, the first steps being taken to bring about his downfall even before Khrushchev had been formally removed from power.[24]

His downfall was accompanied by reassurances from the highest authorities that the same mistakes would not be repeated. At the March 1965 Central Committee plenum Brezhnev declared:

> True science takes nothing on faith; it cannot be the monopoly of particular scientists, and still less of administrators, no matter how much prestige they may have. Unfortunately, we have recently had instances in which people incompetent in science at times took upon themselves the role of arbiters in disputes among scientists, and in so doing hampered their initiative and stood in the way of free, creative discussion of scientific topics.[25]

While primarily an attack on Khrushchev, there was no doubting the damnation of Lysenko as well. The reassurances that Lysenkoism would not be allowed to happen again have been repeated since. One of the most significant was a 1967 *Kommunist* editorial which, while reintroducing the concept of *partiinost'* into science, also attacked scientific 'schools' that had established themselves as scientific monopolies.

> Such a thing happened not long ago with one 'school' in biology. Making use of the conditions of the time, it transferred the struggle with those that thought differently from it out of the sphere of scientific discussion into the field of arbitrary judgements and conclusions, doing enormous harm to the development of branches of biology with very great potential. We must never forget that sad experience, in order to ensure that it never happens again.[26]

The significance of this statement is that at a time when the authorities were clearly preparing for a crackdown on dissident scientists, they were at pains to reassure scientists that the crackdown should not be seen as a restoration of the conditions that gave rise to Lysenkoism. Some scientific disciplines continue to be damned, such as Freudian psychology and sociobiology,[27] although these are disciplines which also are highly controversial in Western societies (not that that justifies their persecution in the Soviet Union). Some Western commentators have tried to show that plate tectonics has found itself in recent years in the same situation as genetics did under Lysenko, the claim being that V. V. Belousov, although no longer relying on appeals to ideology, still uses his bureaucratic position to repress the 'mobilist' theories of continental drift and plate tectonics, in order to protect his 'fixist' theories.[28] However these commentators would seem to be exaggerating the situation; indeed the case seems to be instructive in that it shows us what the 'rules of the game' now are in Soviet academic debates, including ones as bitter as this one seems to be. It is therefore worth devoting a little attention to it.

Certainly plate techtonics took longer to catch on in the Soviet Union than in the West, and Belousov has been outspoken in his criticism of the theory. There are hints that there might have been something murky in the past. In ostensibly a defence of Soviet science two 'mobilists', A. Monin and O. Sorokhtin, write that 'the appearance in geology of the new theory [of plate tectonics] led to a profound transformation of classical geological ideas and produced a genuine revolution in the Earth Sciences. In its significance for geology the new theory could without any exaggeration be compared to quantum mechanics in nuclear physics, molecular genetics in biology and cybernetics in engineering.' One wonders whether the choice of three branches of science with such controversial histories is accidental.[29]

However, in the period since plate tectonics became popular in the West – really only since the first half of the 1960s – there is no evidence of 'mobilism' being repressed. Its adherents have had no obvious trouble getting into the Academy (A. V. Peive, corresponding member, 1958, and full member, 1964;[30] P. N. Kropotkin, corresponding member, 1966; V. E. Khain, corresponding member, 1966), while 'fixist' ideas were openly criticized at an early stage.[31] Indeed an open debate has been conducted in which a policy of balance seems to have been maintained. A 'mobilist' article is usually followed by a 'fixist' reply.[32] It is possible that Belousov uses his bureaucratic positions, mainly in international organisations and as chairman of the Interbranch Geophysics Committee and the Academy's Scientific Council on the Earth's Crust and Upper Mantle,[33] to direct research into areas that will support his theories. R. M. Wood suggests that Soviet research drilling programmes have been so manipulated.[34] But to speak of a 'monopoly' is to exaggerate.[35] Science is so specialised and there are consequently so many disciplines, institutes, commissions and councils, that some degree of 'pluralism' is inevitable. Thus the 'mobilists' have their 'schools' in the Academy's Geology Institute under Peive and Monin's Institute of Oceanology. A true monopoly could be established and maintained only with the help of the very highest scientific and political authorities. In the post-Lysenko era it is unlikely that such help will be forthcoming.

Another contemporary case is perhaps even more instructive, since it includes a major ideological component, and indeed seems to represent an example of a scientist losing out because of his excessive appeals to ideological orthodoxy. This is the case of the dismissal of Nikolai Dubinin as director of the Institute of General Genetics. Dubinin is a geneticist of long standing who has twice been dismissed in anti-genetics campaigns, once following the infamous 1948 conference of the Academy of Agricultural Sciences and once in 1959, when he fell victim to a characteristic piece of Khrushchevian pique.[36] However despite these apparently impressive credentials he has long been unpopular among other geneticists and their supporters for his alleged lack of scruples. It is claimed that he was always prepared, if not to submit to Lysenko, nevertheless to use opportunities created by Lysenko for his own benefit. He is considered to have played a particularly reprehensible role in the condemnation and dismissal of Kol'tsov as director of the Institute of Experimental Biology in the late 1930s.[37] There has therefore long been considerable antipathy towards him, most clearly expressed in the memoirs of Raisa Berg.[38] The antipathy boiled over in 1980 after he published an article in *Kommunist* attacking in strong ideological terms such established and powerful geneticists as D. K. Beliaev and the deceased B. L. Astaurov. The criticism was in terms of the nature–nurture debate, with Dubinin taking the unexpected position for a geneticist of putting the stress on social environment rather than heredity in the formation of the human personality.[39]

A considerable debate continued. As is so often the case the actual positions of the protagonists were not all that far apart, and one suspects a lot of the debate was based on personal antipathy. Dubinin's chief opponent was the mathematician and amateur philosopher A. D. Aleksandrov. Aleksandrov has always been willing to engage in semi-ideological polemics,[40] and was also an opponent of Lysenkoism in all its forms.[41] He used the General Meeting of the Academy of Sciences in November 1980 to launch a bitter attack on Dubinin.[42] We have no details on the bureaucratic politics that followed, but it was reported at the end of 1981 that Dubinin had been removed from his post as director of the Institute of General Genetics.[43] The genetics 'establishment' had triumphed, although it did so apparently with the support of the highest party leadership. In the issue of *Vestnik*

Akademii nauk that announced Dubinin's dismissal, a report was published of a conference on philosophy and the natural sciences. The nature–nurture issue was discussed, with most participants calling for a compromise. However Brezhnev's statement to the 26th Party Congress, giving social scientists and philosophers the task of 'overcoming recidivists of scholastic theorising', was also quoted.[44] Brezhnev's message seemed to be that Marxist–Leninist theorising should not be used in scientific polemics, with the approbation presumably being directed primarily against Dubinin with his appeals to Marxist orthodoxy.[45]

This of course was the most striking thing about the debate and Dubinin's defeat, that his position was the more orthodox one. Marxism has always disapproved of any tendencies towards biological reductionism,[46] and Lysenko's destruction of genetics turned that disapproval into dogma, one retained by most ideologists.[47] Thus in this case the scientists, with the support of the top political leadership, were able to get rid of an old enemy by overthrowing ideological orthodoxy and asserting the dominance of a pure genetics approach.[48]

After the death of Brezhnev, Chernenko clearly supported Dubinin's position, claiming at the June 1983 Central Committee plenum that 'one can hardly recognize as scientific conceptions those which explain such human qualities as, for example, honesty, bravery and orderliness by the presence of 'positive' genes and which specifically deny that these qualities are formed by the social environment.' Further, in contrast to Brezhnev's criticism of 'scholastic theorising', he put the stress on 'ideological (*mirovozzrencheskaia*) precision, or if you like, methodological discipline of thought.'[49]

Despite this, there were no signs under Chernenko of a Dubinin comeback. Indeed even the Academy's ideologists, like Fedoseev, are determined to put the matter behind them, for they fear too great an interest in scientific matters from party ideologists and leaders.[50]

The case suggests that in some controversial branches, such as biology, both ideology and the political leadership can be called on to play a role. However the lesson seems to be that a scientist will make such an appeal only at some risk to himself. In this case the political leadership seems to have gone out of its way not to antagonise the scientific community. In general it can be said that the party authorities have kept their promise not to allow the emergence of another Lysenko.

Even in the social sciences, the period since the death of Stalin and particularly since the fall of Khrushchev has seen a considerable broadening of permitted fields of research. The boundaries imposed by Marxism–Leninism still exist and severely limit the social sciences, while they are also still strongly affected by changes in ideological line among the top leadership. However, they also exhibit great resilience in the face of considerable ideological pressure. This is well illustrated in the cases of mathematical economics and sociology.

To begin with economics, under Stalin it had been, not surprisingly, one of the most ideological disciplines. Particularly in the academic sphere, it involved little more than the simplest possible repetition of the latest thesis expressed by Stalin. The economists who controlled the discipline had little understanding of technical applied economics, and their inbred inclination was to see Stalinist-type control of the economy as the only model ideologically permissible for the Soviet Union. The search for more rational methods of economic management after the death of Stalin exposed to the full the inadequacies of these economists, and criticism of traditional economics, particularly in the Institute of Economics, became widespread.[51]

But the Stalinists in the economics establishment wielded great power, and progress against them was slow. In 1961 Keldysh's exasperation with the economists was clear as he told the Academy's General Meeting that a Department of Economics could not be established in the Academy while disputes over the responsibilities of the projected department continued.[52] Nevertheless, the next year the department was established, with the party activist and old style economist A. A. Arzumanian as academician secretary. In 1963 the Central Economics Mathematical Institute (*TsEMI*), with its emphasis on the technical economics that was anathema to the traditional ideological economists, was established.[53]

However the traditionalists, although losing their monopoly, were far from defeated, and mathematics was still looked on with suspicion in the Institute of Economics.[54] At the height of the ideological relaxation in 1965 the new economists really began to make their presence felt. In that year A. M. Rumiantsev, a confirmed 'liberal',[55] was made acting academician secretary of the Academy's Department of Economics and Gatovsky was appointed director of the Institute of Economics. Gatovsky was a traditional economist, but perhaps because of his persecution at the hands of the economics establishment during the ideological campaign of 1951, or for the more cynical reasons put forward by Belitsky, he supported the 'liberals'.[56]

One might have expected that with the crackdown on political dissidence from 1967 the old guard economists would have succeeded in reasserting their strength. They had certainly not given up the struggle in the intervening years. Rumiantsev could not get his appointment as academician secretary in the Department of Economics confirmed, and at seminars and conferences the traditionalists, led by Iakov Kronrod, continued to attack the mathematical trend in economics.[57] At the 1966 Academy elections, Rumiantsev's election to full membership was offset by the elevation of Khachaturov, the leader of the conservatives.[58]

But the support of not only Keldysh and the Academy's presidium but also the party leadership was given to the mathematical economists. Thus Fedorenko, the director of *TsEMI*, increasingly became the spokesman for the economic community as a whole, while Rumiantsev was confirmed as academician secretary of the Department of Economics in 1967. He had to give up the position soon after, in order to take up a vice-presidency of the Academy of Sciences and in 1969 chairmanship of the Academy's Social Sciences Sector. His place was taken by Khachaturov, an ominous sign, particularly for the market economists in the Institute of Economics. Indeed the events of 1968 in Czechoslovakia ensured their downfall. The market and mathematical economists had much in common, and the defeat of the former meant that Fedorenko had to engage in some 'trimming' of his position.[59] Nevertheless, on the whole the mathematical economists survived the crisis well and indeed strengthened their position.

Developments in sociology followed a similar pattern. Sociology had been destroyed by Stalin in the late 1920s, but within a few years of his death moves were afoot for its rehabilitation. Zemtsov claims that in 1955 G. V. Osipov approached Khrushchev, with the support of the philosopher F. Konstantinov, then head of the Agitprop Department of the Central Committee, with a proposal for a sociological study of the reasons for people leaving the Far North and Virgin Lands.[60] That study was followed by others, and sociology built itself some solid support among regional party leaders for its utility in social and economic planning. It further established its legitimacy by the sending of delegates to foreign conferences.[61]

However, the new discipline faced opposition from the traditional philosophers, with their control of the Department of Philosophy and Linguistics of the Academy and the journal *Voprosy filosofii*. The only organisational success sociology had was the setting up in 1958 of the Soviet Sociological Association.[62] Despite public support for sociology from such party leaders as Kosygin and Il'ichev,[63] the discipline could

not establish a central organisation for itself until, seemingly paradox-
ically, 1968. Despite the opposition of the conservatives and the ugly
tenor of the time, the Institute for Concrete Sociological Research
(*IKSI*) was set up within the Academy of Sciences, according to
Zemtsov following a Politburo decision. It was made directly subordinate to the Academy's presidium, meaning that it was not subject to
interference from the Department of Philosophy and Law.[64] The busy
Rumiantsev was made director.[65] Again according to Zemtsov,
Rumiantsev managed to free himself of the control of the unfriendly
Philosophy Sector of the Central Committee Science Department,[66]
while his appointment in 1969 as chairman of the Social Sciences
Sector of the Academy gave him control of Academy research in
sociology as well as economics. In the middle of 1968 the conservative
M. V. Mitin was replaced by I. T. Frolov as chief editor of the journal
Voprosy filosofii, a change which led to an extensive and positive
coverage of sociology in the influential journal.[67] It looked again as
if the conservatives had suffered a defeat at a time when one might
have expected a victory. The lesson is, of course, that one should by
no means assume that the traditional ideologists represent or have
the support of the party even at times of ideological severity.

However time was to show that the mathematical economists and
sociologists had not won a final or complete victory. While the
sociologists had solid support among sections of the party apparatus,
many of their findings were becoming increasingly embarrassing, show-
ing up shortcomings in the system as a whole and in the work of
powerful individuals in particular. When the major attack came it was
ostensibly over ideological matters. Despite the formation of *IKSI*,
sociology was not properly established in universities and sociologists
had not been able to get a suitable textbook cleared for teaching
purposes. So Iu. A. Levada, the secretary of the party organisation
of *IKSI*, had a series of his lectures, given at Moscow State University,
published in the Bulletin of the Soviet Sociological Association. They
were apparently short of the theoretical Leninist apparatus considered
essential for materials used in teaching, and the institute's opponents
went on the attack.[68]

Demichev, formerly the Stalinist Central Committee secretary for
ideology but now Minister for Culture, had been behaving in a rather
liberal fashion for the previous few years. But now he came out
against Levada, as did Konstantinov, now speaking on behalf of the
Department of Philosophy of the Academy. A considerable struggle,

described by Zemtsov, ensued.[69] Rumiantsev tried at first to keep aloof, but was eventually forced to defend members of his staff and to demand a lessening of party and police pressure on the institute. He was unsuccessful. In early 1972 his deputy director, G. Osipov, was removed as president of the Sociological Association and deputy director of *IKSI*, and the Academy's Scientific Council for Concrete Sociological Research was merged into Gvishiani's Council on the Economic Problems of Scientific–Technical Progress.[70] Mikhail Rutkevich, a physicist by training with little sociological research to his credit, became president of the Association, and then in July 1972 it was reported that Rutkevich had replaced Rumiantsev as director of *IKSI*. Rumiantsev was eventually to lose all his positions, plus his Central Committee membership. Soon after, the institute was renamed the Institute of Sociological Research (*ISI*), the removal of the word 'concrete' indicating the dissatisfaction of the authorities with the lack of Marxist–Leninist theory in the institute's work. Rutkevich presided over a ruthless purge of the institute which saw about 40 per cent of its staff lose their jobs.[71]

These events in sociology had their repercussions in economics. In the early 1970s the traditional economists in the Institute of Economics went back on the offensive. Rumiantsev, with his problems in sociology, was apparently unable to protect the mathematical economists in that institute, and in 1972, following a Central Committee decree on the primary party organisation of the institute, containing criticisms of the choice of research projects in the institute and theoretical mistakes in some of its publications,[72] Gatovsky was removed as director and replaced by the extraordinarily unqualified Kapustin, a change engineered according to Belitsky by the pillar of the economics establishment, Tigran Khachaturov, as the first stage in his attack on Fedorenko. Despite the fact that the Central Committee decree referred critically to the lack of attention to mathematical economics in the institute, it is said that the change in leadership led to a purge of mathematical economists.[73]

One cannot deny the ideological significance of these developments in economics and sociology. The new successes of the traditional economists and philosophers surely reflected a change in policy among the top leadership. However, in both cases the most important factor could be seen as the downfall of an overextended and exposed highflyer. Perhaps the most interesting thing is that *TsEMI* continued to wield great influence and its director, Fedorenko, took over from

Rumiantsev as the head of the economics establishment. He replaced Khachaturov as academician secretary of the Economics Department of the Academy in 1971.

Sociology also made a rapid comeback. In 1976 Rutkevich was replaced by an economist, Timon Riabushkin, as director of *ISI*. Robert Toth puts his departure down to the open complaints of party bureaucrats that the institute was not worth consulting, its work was so poor.[74] From 1972 the Academy of Sciences had made clear its dissatisfaction with the attacks on Soviet sociology. Even at the presidium meeting in 1972 that formalised the restructuring of the institute, the curious statement was made that one of *IKSI*'s shortcomings had been its lack of use of statistics. One of Rutkevich's criticisms of the work of the institute before his assumption of power was its *anketomaniia* (mania for questionnaires) and excessive use of mathematics.[75] In 1975 a rival institute, the Institute of Socio-economic Research, was set up by the Academy in Leningrad, headed by the economist Gely Cherkasov.[76] When the Academy commission, whose investigation of *ISI* apparently led to Rutkevich's dismissal, reported to the presidium, one of the main complaints was that the high turnover of staff, i.e. the purge, and the lack of stability in management personnel, had had a negative influence on research work. Petr Fedoseev, the Academy's chief ideologist, agreed.[77] Since 1976 the reliance of the institute on the party apparatus has not been reduced, with a large proportion of its work still being commissioned directly by regional and local party organs.[78] However independent research, including such sensitive areas as the class structure of Soviet society and the meaning and nature of communist society, has continued and been published in the institute's own journal *Sotsiologicheskie issledovaniia*, and in *Voprosy filosofii*.

Since the death of Brezhnev there have again been signs of difficulties for *TsEMI* and *ISI*. At the June 1983 Central Committee plenum, an ominously conservative speech by Chernenko contained criticism of both institutes.[79] *TsEMI* was subsequently savaged by the political economists at a meeting of the Academy presidium, with the presidium formally censuring the work of the Department of Economics, led by Fedorenko.[80] At the March 1985 General Meeting of the Academy Fedorenko was removed as academician secretary of the Department of Economics. Rumour has it that he has also retired as director of *TsEMI*, and that the institute is to be split into two.

Nevertheless, at the June 1983 plenum a new Centre for Public Opinion Research was ordered to be set up within *ISI*, surely an

indicator of some degree of trust in the institute,[81] while the future
of mathematical economics seems assured within the new structure
of *TsEMI*. Curiously, the Institute of Economics has also recently
been the subject of a critical Central Committee decree.[82]

One gets the impression that the latest criticisms of sociology and
mathematical economics are the result of dissatisfaction more with
the practical results of their work, rather than with their ideological
positions. The party has come to rely on useable social science disciplines. Thus, ideology, while still a restraining influence, is less likely
to be a cause of complaint. Even the political economists at the
Academy presidium meeting mentioned above referred sneeringly to
the lack of practical results produced by *TsEMI* rather than to ideological errors.[83] Advice received from social scientists is expected to be
within the ideological framework of the existing Soviet system, but
this is for practical reasons as much as for ideological control – it has
to be advice that can be used in present Soviet conditions.[84]

In conclusion, the formal ideological status of science has improved,
as a direct productive force and as part of the STR. Examples from
both the natural and the social sciences suggest that its informal
ideological position has also been strengthened. The authorities are
determined to avoid a repeat of Lysenkoism in the natural sciences,
while in the social sciences the party authorities appear to understand
that effective policy-making requires some broadening of horizons.
It would be wrong to think that the party has abandoned its right to
exercise ideological control of Soviet science. As long as the party
holds to the fundamental Leninist principle of its leading role, that
it is the representative of the whole people and therefore the embod-
iment of social progress, no further ideological or theoretical legitima-
tion is needed of its right to determine which science is socially progressive and which is not. The fact that such a right is used to a far
lesser extent than under Stalin does not mean that it does not exist.
In the social sciences it is still clearly a major factor. In the natural
sciences it is now perhaps a right so rarely used that it has become
of little significance.

AN 'ETHOS OF SCIENCE'

Some would go so far as to claim that Marxism–Leninism is not a
negative factor in Soviet science, but that it indeed plays a positive
role. This is not a reference to the supposed beneficial use made of

the principles of dialectical materialism in scientific research, although Loren Graham has written a most interesting book on such a theme.[85] Rather, the reference is to the inherent and deep commitment to science that many Western, and of course all Soviet, commentators see in Marxism and Marxism–Leninism. Marx granted a major role to science, through its transformation of industry, in the progress of mankind.[86] Lenin demonstrated his practical commitment to Marx's view with vigorous efforts to promote scientific research in the early difficult years of the Soviet regime.[87] As an example of Western views, Loren Graham states that the Soviet Union 'is a nation with an explicit commitment to science, including a value system and a philosophical world view based on science, which is unmatched in intensity by any other nation in the world'.[88] R. F. Miller gives this finding practical significance by stating:

> The culture, or perhaps more accurately, the 'cult' of science and technology in the USSR does appear to generate a significant impetus toward the concentration of social resources on these crucial elements of national power in the modern world.[89]

Some Western writers go so far as to claim that the Soviet regime's legitimacy rests on this commitment and the scientific progress resulting from it.[90]

Certainly one can find support for such a commitment in the special attention that all past Soviet leaders have devoted to science, technology and scientists,[91] in consistently high levels of science funding, and in an attempt at least in recent years by the party leadership to organise the Soviet system of industrial administration on the basis of a commitment to science and technology.[92]

However, before we allow ourselves to be too impressed by the benefits of this commitment, some other aspects of it need to be examined.

Western 'culture of science'

It would be wrong to think that Marxism and Marxism–Leninism are unique in being committed to science. Marx's belief in science was very much derived from the capitalist ethic of the time – indeed Marx believed that the capitalists would solve the problem of technology and it would be left to the workers to enjoy the fruits after the revolution.[93] Soviet leaders and academics have also shown themselves constantly impressed by Western methods and achievements. There is probably just as much faith in science as a panacea for economic and social ills in the West as in the Soviet Union, and about the same willingness to fund research. The various 'anti-science' movements that have appeared since the 1960s have made only the very tiniest of dents in that Western 'culture of science'. Further, Soviet efforts provide a constant impetus to Western governments for the maintenance and strengthening of the Western commitment.

Domestic versus foreign science

A Bolshevik commitment to science does not necessarily mean a commitment to domestic science. The Bolsheviks, right from the beginning, were faced with the inescapable problem that their domestic science was not even close to the level required for the modernisation of the Soviet economy. Lenin found it easy, ideologically and practically, to justify looking abroad for technology at the same time as devoting attention and resources to domestic science, both pure and applied. But with the development of the Soviet economy beyond the virtually hand-to-mouth methods of the Civil War, War Communism and NEP periods, more resolute strategic decisions had to be made. Stalin's decision was to devote all possible resources to the most rapid possible industrialisation. This made inevitable the large-scale import of foreign technology, often on a turn-key basis. While impressive resources were still devoted to domestic research, there were three features of the science strategy of the time that have had important long-term consequences. Firstly, Soviet applied research was allocated an ancillary role servicing imported technology;[94] secondly, this work was done in central institutes covering a whole industry – enterprises had and still have very little R&D capacity of their own;[95] and theoretical research, despite ideological statements on its necessary links with production, in fact developed in rather isolated circumstances. These gaps between pure research, applied research and production continue to be serious problems for Soviet R&D.

Stalin maintained an import-based R&D strategy during the 1930s, despite an ideological stance which emphasised the inherent superiority of Soviet science over 'bourgeois pseudoscience'. This strategy was abandoned in the postwar years, and was accompanied by a particularly vicious 'anti-foreign science' campaign. It was only after the death of Stalin that Soviet science was recognised again as being part, and often a lagging part, of the international scientific community. This was of benefit to scientists in that it opened the way for them to establish contacts with foreign scientists and genuinely involve themselves in international scientific communication. However it also made possible, ideologically and practically, an extensive use of technology imports. This was most noticeably put into effect during the period of *detente* in the early 1970s. The economic side of *detente* included claims of Soviet participation in the international industrial and scientific division of labour, and imports of technology reached a high level. Some scientists saw this as a mixed blessing, obviously fearing that too great a reliance on imported technology would lead to a neglect or 1930s-style distortion of Soviet science. This fear was most openly and insistently expressed by A. P. Aleksandrov, the president of the Academy of Sciences from 1974.[96]

One suspects that for reasons of pride and strategy the Soviet leadership will always yearn for scientific and technical autarchy. To this extent Aleksandrov's fears are exaggerated. Nevertheless, historically it has been demonstrated that the leadership is prepared to go abroad for its technology when the international situation allows it and the economic effect promises to be great.

'Anti-science' attitudes in the USSR

A 'production' ethos

The course of Soviet history so far has seen an emphasis on the quantity of production at the expense of quality, on an engineering approach at the expense of considerations of economic and social efficiency, and on the fulfilment of plan indicators for gross output as the only important measure of job performance. The leadership has striven manfully in recent years to change this approach, but one suspects that its failure is due as much to its own psychological barriers to change as to bureaucratic problems. While the industrial bureaucracies have their own selfish reasons for maintaining the present system, the leadership is perhaps psychologically unable to break the habits of long careers in party and economic management. (There is also a realistic recognition on its part that any loosening of the planning grip on the industrial bureaucracies would most likely lead to a drop in output without any compensating improvement in quality and techniques.) The effect on R&D is well-known – innovation is something to be avoided because it threatens plan fulfilment.

Another potentially damaging aspect of the Soviet commitment to production is that science is given a value only in so far as it contributes to production. Despite some allowance being made for the cultural and anti-religious importance of science, the connection with production has always been at the basis of the Marxist attitude to science. Marx saw the significance of the natural sciences in their transformation of industry, not in any role they might play in general enlightenment.[97] Science in the Soviet Union has never been allowed to justify its existence simply by referring to increases in human knowledge or improvements in man's understanding of his place in the world. Particularly as a direct productive force science has a legitimate existence only within a symbiotic relationship with production. The Soviet leadership has generally interpreted this requirement quite liberally, and esoteric theoretical research has been allowed to proceed. Nevertheless, the tension between theoretical and applied research in the Academy of Sciences, the major centre for Soviet fundamental research, is a constant factor in the Soviet politics of science. This was most evident during the 1930s,[98] and in the second half of the 1950s, with the debate between the theoreticians and applied scientists in the Academy. But the tension is still there today.[99]

Anti-intellectualism
Despite the psychological commitment of the leadership to gross output and, one suspects, a feeling among production engineers that scientists are 'ivory tower academics', anti-intellectualism does not seem to be a major phenomenon in Soviet society. It is true that early in the Soviet regime the trade union debates over the power of 'bourgeois' managers and specialists versus trade union democracy had an important class-conflict component, but the debate was resolved in favour of the 'bourgeoisie'.[100] Later in the 1920s Stalin used working-class resentment of 'bourgeois specialists' for his own political purposes, but again the bourgeois specialists were eventually reestablished in positions of power.[101]

In recent years the leadership is said to be following a policy of a type of 'social contract' with the working class, at the expense of the intelligentsia. This is evident in pro-working class party recruitment policies,[102] the narrowing of income differentials,[103] as well as the continuing ideological and propaganda stress on the ideological superiority of the working class, including the status of the intelligentsia as a stratum, not a class.[104]

However this policy does not seem to have been applied firmly enough to have removed the material advantage of the intelligentsia,[105] or the status of intelligentsia jobs, particularly research jobs. Soviet surveys show without exception the high desirability of careers in research for children leaving school, their parents and the population as a whole.[106]

However it should be noted that in recent years concern has been expressed about a reduced interest among school leavers in scientific research, with a significant movement towards medicine, the social sciences, law and journalism.[107] It is interesting to speculate why this might be. It might be simply one of those inexplicable vagaries of fashion, but there do seem to be more concrete explanations. It might be a popular reaction against the natural sciences which are seen as having provided the basis for the threatened destruction of mankind; it might be that with the greater role of economists, sociologists and lawyers in Soviet policy-making in the last twenty years these disciplines have gained a respectability and attractiveness that they did not have in the past; perhaps the political repression in research institutes since the mid-1960s and the relentless insistence in the Soviet press and scientific literature on the collective, planned and production-oriented nature of modern science has convinced youngsters that scientific research is no longer the last bastion of freedom; it might simply be that with the enormously high recruitment since the mid-1950s a career in science no longer carries with it guaranteed promotion and *propiska* (a residence permit) in one of the major cities.

There are in the Soviet Union some of the symptoms of a worldwide anti-science movement, the significance of which even in the West has perhaps been exaggerated. The conservation movement is strong in the Soviet Union, but does not seem to have the anti-technology, anti-progress connotations that it has for some in the West. Certainly there seems to be no feeling that scientists are to blame for environmental problems, or that science and the scientists are the allies of industry and the state in the despoliation of nature. Indeed the leading figures in the major conservation movements have been scientists, the Lake Baikal campaign being the best known example.[108]

There has been a movement in literature and philosophy of literature opposed to progress based on science and technology and the dehumanisation they bring. This trend is in striking contrast to traditional socialist realism in which science and technology invariably play a positive role. One of the pioneers of the trend was Vladimir Tendriakov. In his 1962 novel *Short Circuit* he demonstrated an ambiguous attitude towards technology and progress. While they were not rejected outright, he suggested that if they were not to do harm they had to be supported by spiritual values that were little in evidence in the modern world. Later works, such as *Apostolic Mission* (1969), showed more strongly the hero's loss of faith in the capacity of science to reveal the ultimate truth of the universe.[109]

The 'village prose' writers then went further and developed a whole world view based on the traditional, simple ways of Russian village life. According to Geoffrey Hosking, a member of the village prose camp such as the critic Lev Anninsky in fact went further than simply an appeal to traditional Russian values, and asserted '*personal* values in general against a technology and society which were threatening to become impersonal'.[110]

A literary philosopher such as Iu. N. Davydov, although arriving at somewhat different philosophical conclusions, is also concerned with the loss of human values involved in the deification of science and knowledge. In such circumstances 'the moral aspect of the under-standing of truth, and the very use of philosophy has become clearly irrelevant.'[111] This particular debate is centred around Dostoevsky's ethics. Dostoevsky's writings were, of course, in many ways a reaction against the extreme reductionism of the Russian positivists of the 1860s. These positivists, such as Chernyshevsky, had a considerable influence on Russian Marxism.[112] Geoffrey Hosking describes the position of Lev Anninsky as specifically a rejection of the post-Chernyshevskian positivist tradition, and includes Lenin in that tradition.[113]

The success of village prose might indicate that it represents some popular feeling against technology and modernity. But it is highly unlikely that such views, relatively widespread in literature and literary criticism, have had any effect on the actual conduct of science. They are only barely tolerated by the Soviet ideological authorities – indeed, under Chernenko it seemed that a relatively relaxed attitude to the village prose school could be accompanied by attacks on others who write ill of the scientific community and its values.[114]

It should be added that the non-literary theorists of the STR are not unaware of the dangers of amoral 'technocratism' and alienation inherent in rapid technological advance. Such aspects of the STR have received considerable attention in recent years,[115] although the solution to the problem is all too often a simple declaration that the party or socialist society will of themselves provide the necessary injection of spiritual values into the STR.

The Soviets seem to have at least diluted one traditional source of popular anti-intellectualism (very evident during the anti-cosmopoli-tan campaign of the late 1940s and early 1950s) by creating a new non-Jewish intelligentsia, and indeed by making that new intelligentsia one of the main centres of modern Soviet anti-semitism.[116]

Attraction to 'false science'[117]

False science, in its two senses of unscientific methods being used in accepted disciplines and research in disciplines not generally accepted by the scientific community, is probably now no more common in the Soviet Union than in the West. But it tends to invite more attention in the Soviet Union, firstly, because there any non-conformity stands out, and secondly, because any such research must at the very least be tolerated by authorities who rarely show tolerance of things of which they disapprove.

I have already mentioned Lysenkoism, the best-known example of false science of the first type. It is now generally considered that Lysenko's success was not because of the greater compatibility of his biology to the ideological tenets of Marxism–Leninism, but because Lysenko made irresistible promises of rapid agricultural growth that could not be matched by the more sober and necessarily patient geneticists. The question then becomes, Why did the authorities fall for Lysenko's promises even at the expense of common sense? A number of Western commentators agree that the leadership found itself under the spell of a practical messianism which declares anything to be possible in a socialist society. Eric Ashby, writing in 1947, talks of 'a reckless optimism, a bravado almost, in large-scale technical research and in far-distant exploration', an attitude for which Lysenko was perfectly fitted.[118] R. M. Mills describes how the leadership's caution about Lysenko was overcome:

> That reserve grew out of the respect for science and specialists that was part of the Bolshevik heritage, but it became submerged in another part of the same heritage, the wave of obligatory optimism in the face of enormous tasks, in this case the task of achieving rapid progress in agriculture.[119]

This attitude is still to some extent in evidence, not just in the leadership's insistence on the practical utility of all research, but in the role the party still assumes as a 'vanguard' in all areas, willing to listen to and adopt radical ideas, overcoming in the process the conservatism and inertia of the established bureaucracies. It seems that regional and local party workers are particularly susceptible to the claims of at best those with ideas unacceptable to the scientific establishment, and at worst scientific cranks and frauds.[120]

Of a somewhat different nature is Soviet research into such phenomena as telepathy, ESP, parapsychology and UFOs. One assumes that the party often allows such work to proceed through ignorance or indifference towards what some of the more way-out scientists are up to.

However, the Russian people, for whatever reasons, seem to have a strong and unsceptical attitude towards such phenomena, an attitude, which rumour has it, has been shared by Soviet leaders. It was 'common knowledge' that towards the end of his life Brezhnev used the services of the faith-healer, Dzhuna.[121] That makes the personal attack on Dzhuna by Academician Zel'dovich particularly pointed. In 1981 he criticised the Academy's Department of Philosophy and Law for discussing the philosophical aspects of biophysical phenomena and for inviting Dzhuna along. As Zel'dovich exclaimed: 'Not even the likes of Rasputin got into meetings of the Russian Academy!' (One is unsure whether Zel'dovich is implying that Dzhuna played the same role in Brezhnev's circle as Rasputin did in Nicholas II's court.) Aleksandrov, the president of the Academy, backed Zel'dovich, and Anatoly Egorov, the academician-secretary of the Philosophy Department, announced that one of the Department's corresponding members had been informed of 'the impermissibility of replacing the serious examination of scientific questions with cheap sensations which interfere with the development of science'.[122]

This case was unusal only in that it could be interpreted as a criticism of Brezhnev. Academicians regularly publicly express irritation that such research is being done in the Academy, or that people believe it is being done.[123] One suspects that it is more a matter of irritation than anything else. While a phenomenon such as Lysenkoism cannot be ignored – and one feels it is still a very relevant issue in party–science relations – false science of the second type would not seem to constitute any real threat to mainstream science in the Soviet Union.

I would see the production ethos of traditional Soviet economic management as the most important qualification to a Soviet scientific ethos, which makes some recent developments to be described later in this study particularly important. However, I am prepared to accept that virtually throughout the history of the Soviet Union the leadership has displayed a particular commitment to science and technology, a recognition that scientific development is an essential aspect of national development and therefore deserving special attention. This has on occasions led to an overexuberant attachment to particular scientific 'schools', an attachment made even stronger by the ideological manipulations of those with particular interests to advance. Now that the ideological factor has declined in importance, the commitment to science is perhaps less likely to produce distorted results.

BUREAUCRATIC CORRUPTION

Many would say that so far one major negative factor in the Soviet culture of science has been ignored. They would agree that the cynicism of the party leadership reduces the significance of Marxist–Leninist ideology *per se*, and might agree that the Bolsheviks do have a commitment to science. However, they would claim that two phenomena exist which corrupt the atmosphere in which Soviet science operates and guarantee that the political sphere can have only a negative effect on science.

Conservative mindset

Firstly, the claim is made that a rigid and essentially conservative mindset is produced by Soviet education and indoctrination. This is seen in the lack of truly pioneering research produced in the mature Soviet system,[124] and the lack of analytical skills displayed by even those social scientists who have escaped the rigours of censorship by emigrating to the West.[125]

It is a generalisation the validity of which is impossible to determine – measurements of quality are difficult and there are many other factors to be controlled for. Soviet theoretical research has had its successes,[126] while I have been impressed by the analytical abilities of many Soviet emigres once they have become familiar with current Western social theory. This would suggest that any apparent lack of analytical ability is more a result of isolation and the dangers of engaging in genuine analysis while still in the Soviet Union. These are two very important features of Soviet science, but it should not be thought that they necessarily inflict irreversible damage on the intellects of Soviet scientists.

It is possible to limit the argument to those who run science – either the political leadership of the country or the leaders of the scientific hierarchies.[127] These people, ossified by years in the bureaucracy, have neither the imagination nor the courage to support truly original research. I will devote my attention to the scientific leadership later. As for the political leadership, one could argue that the Bolshevik leadership, as mentioned above, has long had a fascination for big new ideas. Not even the most recent grey figures at the head of the party have managed to eliminate totally Bolshevik 'voluntarism'. The Northern and Siberian Rivers projects demonstrate the validity of such a proposition.[128] Soviet science, particularly when encouraged by the party, has often tended to be too radical, too willing to take up new ideas. Lysenkoism is a classic example. It was one of Lysenko's attractions for the party that he was not an established scientific figure, but an outsider struggling against the scientific establishment.[129] The history of polywater is less dramatic and evil but also instructive.[130]

The question of the influence of Soviet education and indoctrination on individuals' capacity for creative analytical thought is a complex and dangerous one. My feeling is that the history of Soviet science and culture suggests that one should be very careful of exaggeration.

The control ethos

The second claim that is made is that, far stronger than any 'scientific ethos', the Soviet Union is dominated by an ethos of political and bureaucratic control. It is derived from two sources: the understanding that tight political control can be assured only by close bureaucratic regulation of all social and economic relations, and secondly, a conviction, going back as far as the mechanist economic theories of Marx, that one hundred per cent economic and social efficiency can be achieved by socialist planning and distribution. Both sources combine to produce an obsession with bureaucratic regulation, with two negative consequences for Soviet science. Firstly, scientific creativity is regulated out of existence. Secondly, room is provided for the operation of unscrupulous and ambitious adventurers, able and willing to play the game of bureaucratic politics.

There are two basic officially-sponsored values in Soviet science that might contribute to the control ethos: *partiinost'* (party-mindedness) and *kollektivnost'* (collectiveness). Each has a number of components.

Partiinost'

Partiinost' implies, firstly, a high degree of political awareness on the part of scientists, both in their work and in their private lives. They are expected to continually strive to broaden their understanding of the relevance of politics and ideology to their science, as well as keep up-to-date with general domestic and international political events.

This is to be achieved through the usual work-place lectures and meetings that are part of life in any Soviet institution,[131] as well as the philosophical and methodological seminars that are specific to scientific institutions. These seminars are primarily intended to discuss the relationship between dialectical materialism and specific scientific disciplines and theories.[132]

Another aspect of political awareness is that scientists are expected to exhibit political conformity and reliability not just in their scientific work but in their private and social lives as well. They, like any other Soviet citizen, are subject to censorship, limitations on the freedom of assembly, meaningless elections, etc.

The second component of *partiinost'* is that scientists exhibit a commitment to the social relevance of their scientific work. This is an aspect of the utilitarian view of science held by the leadership, but here the stress is not just on science policy-making at the top levels, but on the attitudes of every individual scientist. Each scientist is expected to search for research topics that have practical relevance, and to be open to approaches from and to seek out contacts with production personnel.[133] There is, further, the strong implication that researchers, through the planning and project review system, will be forced into working on projects that are seen as beneficial by their superiors.

Kollektivnost'

Firstly, it should be stressed that Soviet *kollektivnost'* is not the same as Robert Merton's 'communalism'. Merton meant that all the achievements of science, including the most brilliant achievements of individual scientists working in isolation, are the collective property of the whole scientific community, and that science advances through the accumulation of this collectively owned knowledge. The Soviet concept means that science is essentially a collective endeavour, that the day of the lone scientist is gone. This is of course a realistic recognition of one of the major trends in science in the twentieth century. However it has far more significance for the Soviets than that. Not only is the inherent superiority of collective enterprise over individual enterprise one of the tenets of socialism, but the collective is one of the fundamental bases of Soviet political and social control. One obtains from the collective not simply one's livelihood, but also housing, pension and social security rights, access to holiday facilities and even to some extent access to food.[134] It is also the scene of a great deal of not very subtle 'peer pressure' to bring about higher labour productivity and conformist behaviour.[135]

The implications of *kollektivnost'* in science are that valuable research can only be done through collective enterprise,[136] and that therefore scientists are just as subject to collective-based controls as any other Soviet citizen. There is then the further implication that science, being a collective-based occupation, is something which must be managed, that is, run by a bureaucracy. Bureaucracy, in the pejorative sense, has long been a feature of Soviet life, and science has always been subject to it. In more recent times a whole discipline of 'science management' (*naukovedenie*) has grown up, which has produced an enormous literature on every aspect of the scientific process – on the size of research units, their regional, spatial and structural arrangement, productivity and bonus schemes, methods of work evaluation, analysis of leadership styles, the study of personal relationships within laboratories, etc. Many of these studies reflect what is undoubtedly an exceedingly bureaucratic and centralised scientific system.

How much do these official values correspond to the values of Soviet scientists, and how much do they affect their work, whether positively or negatively? For information we rely to a large extent on Soviet survey data, which must be treated with some caution. Some information is also available from emigre sources, although one should be equally careful of treating emigres as a suitably random or representative sample of the Soviet scientific community.

Partiinost'

How do Soviet scientists react to the demands of *partiinost'*? Firstly, how politically interested are they? It is generally claimed in the Western and emigre literature that political study is considered by Soviet scientists to be a tedious waste of time,[137] and Soviet evidence can be found that suggests that the Western impression is correct. Particularly with the crackdown on dissident activity in research institutes in the late 1960s some stress was put on the negative consequences of the failure of both leading and rank-and-file personnel to involve themselves in political study.[138] However since then reports have continued to appear of scientists doing their best to avoid political study, with the Central Committee decrees of April 1979 and May 1981 (designed to improve radically political indoctrination throughout the system) not making any obvious difference.[139]

In recent years there has been an enormous increase in the attention paid to philosophical seminars, with this attention resulting in greatly increased participation.[140] But two constant criticisms of these seminars indicate the lack of political interest among scientists – firstly, criticism of the apathy scientists show towards the seminars and the lack of ideological expertise of those running them;[141] and even more common claims that the seminars lose their ideological content and are used by scientists to discuss matters connected purely and narrowly with their current research.[142] This suggests a readiness on the part of scientists to turn political study to their own purposes, if they are not able to avoid it entirely.

What must be more worrying to the authorities than scientists' political apathy is their tendency to become involved in dissident political activity. It seems to be an empirical fact that the scientific and technical intelligentsia make up a large proportion of at least those dissidents known in the West, particularly if we include the Jewish refuseniks.[143]

There are a number of possible reasons for this: the traditional role of the intelligentsia in Russia; scientists are better educated and more used to analytical thought than their fellow citizens;[144] they are the most affected and therefore alienated by political and ideological controls;[145] the difference between the relative freedom they have in their research work and the rigid controls on their private and social lives produces role strain;[146] there is still some solidarity and compactness within the scientific community, with professional values being used to offset threatening political values;[147] and scientists have always had the opportunity to influence policy-making, with this influence being allowed to spread under Khrushchev into politically sensitive areas – specifically the insistence of Kurchatov that the study of radiation genetics be permitted and then his and Sakharov's protests over the resumption of nuclear testing in the late 1950s and early 1960s.[148] I like the theory put forward by Zhores Medvedev that nuclear testing and Lysenkoism had a strong politicising effect on large numbers of senior scientists, that they 'created a unique situation in which the highest scientific support became possible for *political* dissidents who had never belonged to the [scientific] elite.' Medvedev claims that from 1962 senior scientists actively helped the dissident movement, giving financial and material support to *samizdat* and defending those in trouble.[149]

This politicisation of the scientific elite continued after the fall of Khrushchev, despite the removal of atmospheric nuclear testing and Lysenkoism as issues, under the threat of a neo-Stalinist revival led by such conservatives as A. N. Shelepin and S. P. Trapeznikov. The willingness of senior scientists to criticise publicly the activities of the neo-Stalinists encouraged the rank-and-file also to take part in such activities.[150] Thus up until about 1967 the circulation and signing of petitions and protest letters was a widespread phenomenon in Soviet institutes. However the defeat of the neo-Stalinists – clearly signalled by the removal of Shelepin as head of the KGB in 1967 – reduced the willingness of the members of the academic elite to continue their dissident activity. Although they were still prepared to speak out against recurrences of neo-Stalinism,[151] they were not prepared to encourage or assist scientists whose dissidence took on a broader scope.[152]

As is so often the case, the defeat of the hardliners was followed by a crackdown by the former moderates. The minor-league dissidents whose activities were confined to the signing of petitions and protest letters were easily dealt with in the spring of 1968 – the expulsion from the party and dismissal from their jobs of a representative sample of *podpisanty* (letter signers) had an immediate effect.[153] It was already noticeable in 1967 that official enthusiasm for 'science cities', communities built outside the main metropolitan areas purely for scientific research and which had become centres of dissidence, was waning. In that year in Akademgorodok and in early 1968 in Obninsk sociological surveys were conducted which indicated, according to the authorities, low levels of political awareness and commitment. These were used as the justification for a large-scale purge of local leaders and research personnel, the closing down or emasculation of the clubs and newspapers which had been the forums for the dissident intelligentsia, and the eventual dilution of the intelligentsia's domination of the cities by the redrawing of administrative boundaries and the building of factories, resulting in an influx of working-class people.[154] In the late 1960s the primary party organisations of research institutes, and through them the regional party apparatus, became much more involved in personnel and general management.[155]

By the time of the invasion of Czechoslovakia political dissidence as a mass phenomenon among the scientific intelligentsia was already ended. This result had been achieved quite easily and by Soviet standards leniently. Nevertheless, the experience, plus the strong presence of scientists in the continuing hard-core dissident movement, must have made the party authorities very wary of this section of the Soviet population.[156] One assumes that the authorities are not so naive as to believe that their political study programmes are enough to reduce the threat of dissidence among scientists. One should therefore not forget such 'organisational' aspects of *partiinost'* as reduced bonuses, dismissal, party disciplinary measures, etc.

The second aspect of *partiinost'* is that scientists be committed to

the social and economic relevance of their work. There is some survey
evidence of attitudes on this question. A 1966 survey of 2260 graduates
of Estonian secondary schools included in its published results
responses to 'benefit to society' as a motive for adopting a particular
profession. The results of the survey are shown in Table 1.1. As 'very
important' such a motive comes fifth; if we combine 'very important'
and 'average importance' it comes sixth, behind all the values which
could be said to be connected with personal self-fulfilment.

The compiler of the survey reveals in his commentary that about
50 per cent of respondents intending to be scientists thought 'benefit
to society' to be very important, compared to close to 100 per cent
of intending agronomists. He also expressed dissatisfaction with the

Table 1.1   Criteria for choice of profession among 2260 school leavers

The ideal profession must give one the possibility of . . .	Very important %	Average importance %	Not very important %	No answer %
Creating and being original	53.5	33.3	7.7	5.5
Using personal capabilities	75.0	17.5	2.9	4.5
Constant self-improvement and development of outlook	88.0	9.0	0.3	2.5
Being useful to the economy	48.9	37.9	8.4	4.9
Making good money	28.8	58.6	9.3	3.3
Being highly regarded by friends and acquaintances	41.5	43.5	10.4	4.6
Achieving an important place and prestige in society	8.0	37.4	47.4	7.0
Managing people	9.6	27.9	55.9	6.5
Securing a stable and peaceful future for self	20.9	44.5	27.8	6.8

SOURCE: *Voprosy filosofii*, 4/69, p.55.

very high scores for self-fulfilment, since these indicate too strong an attitude of 'study for study's sake', which could lead to problems when the students find themselves in the essentially non-self-fulfilling real world.[157] A. I. Shcherbakov summarises the results of four different surveys of research workers in a table reproduced as Table 1.2. The results are essentially the same as those from the Estonian survey.

Other surveys, at least in the form in which they are available to us, do not specifically include responses on social or economic utility. However they almost invariably reveal that the main value or motivation in scientific research is self-fulfilment or creative expression. Those not working in science for these reasons are working there for so-called 'chance' (*sluchainye*) reasons, presumably meaning good pay, access to housing, no other work, etc. These findings are invariably regarded with distaste by Soviet commentators.[158]

One gets the strong impression from the reporting of these surveys that the Soviet authorities consider a desire for creative fulfilment to be not the most desirable one for scientists, since it leads to unreasonable demands for such fulfilment and consequent disillusionment.

Kollektivnost'
For both ideological and control reasons official Soviet policy puts a high value on the collective spirit – each worker being tied tightly to

Table 1.2   Motive for working in scientific research in four cities

Motive for working in scientific research	Novosibirsk %	Ufa %	Kharkov %	Alma-ata %
Interest in scientific work	67.9	70.5	64.8	81.2
Strengthening role of science in society	9.9	—	—	12.2
Good working conditions	13.6	9.4	6.2	1.1
Desire to make scientific use of one's practical experience	1.2	11.0	1.4	4.5
'Chance' reasons	7.4	9.1	26.2	0.5

SOURCE: A.I. Shcherbakov, *Effektivnost' nauchnoi deiatel'nosti v SSSR. Metodologichesky aspekt*. 'Ekonomika', Moscow, 1982, p. 144.

the 'collective', usually the work place. One does indeed get a sense from much memoir literature of an emotional attachment to the collective that seems strange to a Westerner. It is evident in Raisa Orlova's memoirs of her years in the Institute of Philosophy, Literature and History, and in those of N. Ianevich at the Institute of World Literature.[159] Russian science has a strong 'school' tradition, which often equates with the institute established by or presided over by the founder of the 'school'. There is also considerable rivalry between the major tertiary institutions.[160]

Such a commitment has its positive side. A well-loved and unified institute is more likely to be able to resist outside pressures, including political and bureaucratic pressures. E. P. Velikhov writes with what seems to be genuine pride of the collective spirit in the Kurchatov Institute of Atomic Energy. As a result, management feels confident enough to grant more than usual research freedom to its researchers. Velikhov claims that if you have a good scientist and good collective the scientist should and can be left alone to work in peace.[161] If Zhores Medvedev can be believed, under the directorship of A. P. Aleksandrov the institute has been among the politically more 'liberal'.[162] (However, 'liberality' is not a quality which Aleksandrov has noticeably displayed in recent years, particularly in his attacks on Sakharov.)[163]

But there is an unfortunate side to this type of *kollektivnost'*. Mark Azbel describes the attachment of many Soviet scientists to science, usually embodied in their institute, as 'perverted'. It becomes their supreme value in a world otherwise bereft of emotional and spiritual values, leading them to see the protection of the institute as the supreme moral value. The institute must be protected even at the cost of sacrificing individuals – a situation which in Soviet conditions arises all too often.[164]

One can never be certain how genuine the emotional commitment is – it might in fact be a cover for simple cowardice or a careerist attachment to the party. Mstislav Keldysh, a past president of the Academy of Sciences, displayed this type of *kollektivnost'*, although we cannot know with what motivations. He showed considerable energy in defending science from the likes of Lysenko and was prepared to push for mathematical economics against the opposition of the powerful political economists. But when it came to the *podpisanty* in 1967–8, his defence of science was, in Azbel's terms, 'perverted'. Rather than defending the intellectual freedom of researchers, he claimed that the *podpisanty* had to be stopped, not because they were ideologically objectionable, but because they were placing science in a difficult and dangerous position. The party was doing its best to democratise society and science, 'but the actions of certain scientists interfere with this process. Also contacts with foreign scientists are made more difficult, since now it will be impossible to send many scientists abroad.'[165]

Kollektivnost' can clearly be a dangerous thing, and it is something which is not shared by all. As Linda Lubrano remarks, given scientists' high evaluation of individual creativity, one would expect a low evaluation of *kollektivnost'*. And indeed survey data show such a low evaluation, especially among junior researchers. A Leningrad survey of young researchers found that only just over 25 per cent of respondents considered that work in a collective intensified their creativity.[166] A survey of four institutes in Irkutsk was said to have demonstrated a lack of collective spirit in laboratories.[167] Petr Kapitsa, the physicist, once said that the desire of physicists to avoid having to work collectively had led to an imbalance of theoretical over experimental work.[168]

That the party authorities do not treat such findings lightly can be seen in their claims that it is such "individualism' that led to much of the dissidence of the second half of the 1960s. Petr Fedoseev, chief ideologist of the Academy of Sciences, writes, after a strong attack on 'individualism':

I must draw the attention of the reader to the fact that signs of individualism appear more often among people whose work is essentially of an isolated character. This can be seen in individual cases in science, literature and the arts, where at times individualistic qualities and habits can appear and take root . . . . It is in this area (and often around it) that operate certain people placing themselves outside, and even worse, above society. The boundless, even morbid ambition and, I would say, pathological self-opinionatedness of these people is nothing more than an extreme form of individualism. Those who have been gripped by a mania for power, have declared themselves newly appeared geniuses and claim a special place in society, but not receiving privileges or any satisfaction of their demands, inevitably come into conflict with society, overtly or covertly, silently or stridently.[169]

There seem to be two major reasons why Soviet scientists would want to avoid working in a collective.[170] Firstly, there is the desire to avoid the conflict which apparently occurs within Soviet working collectives.[171] A. I Shcherbakov provides a table summarising the results of a number of surveys on the causes of conflict. It is reproduced in Table 1.3. It speaks poorly of the characters of Soviet researchers,[172] although the findings also reflect rather poorly on laboratory and institute heads.

But a more inescapable problem is the bureaucratic control that goes with working in a collective. This is not the place to go into detail on the planning of scientific research, which in theory and often enough in practice means that each scientist's project must be set out in a central plan which should have an overall practical end, and usually entails fitting his work into the work of others. While it is true that plans are very often in fact simply compilations of the individual projects of individual researchers, particularly if they are senior researchers,[173] their subjection to planning discipline is a common complaint among academics. A recent survey of 7600 research workers in Moscow revealed that only 19.3 per cent thought that planning contributed positively to their work.[174] However, perhaps of greater irritation, because they are less easy to get around,[175] are the strict controls on personnel and wages policies,[176] and the bureaucratic rigmarole involved in obtaining equipment.[177]

Not only do these controls stifle the flexibility and freedom of research, but of course they take up an enormous amount of time. It is estimated that about 20 per cent of the time of even rank-and-file scientists is spent on bureaucratic matters and coping with inefficient organisation, with perhaps a third of the working time of leading researchers being expended on the planning process.[178] The irritation caused by the time taken is increased, firstly by the petty form some of the controls take;[179] secondly, by the inescapable conclusion arrived at after years of experience that the regime is inherently and psychologically unable to relax these bureaucratic controls (probably more so than even ideological controls);[180] and thirdly by the unpleasantness and bitterness that dealing with the Soviet bureaucracy inevitably arouses.[181]

With the increasingly fashionable concern with the application of the methods of the scientific management of labour to scientific research – particularly complex productivity and performance evaluation schemes – things can only become worse. Many of these schemes are designed to introduce greater flexibility into personnel management. In practice they appear to produce even greater bureaucracy and opportunity for greater political control of 'unreliable' researchers.[182]

Table 1.3 Reasons for conflict in research institutes in four cities

Reasons for conflict	Novosibirsk 1964 %	1974 %	Leningrad 1964–7 %	1973 %	Alma-ata 1968 %	Moscow 1969 %
1. *Connected with inadequacies in the management and organisation of labour*	36.8	50.8	45.9	41.8	35.1	33.9
Bureaucratic methods of chief	12.9	24.8	14.1	6.2	14.3	12.9
Scientific disagreements between chief and researchers	—	3.0	5.0	5.2	0.5	—
Lack of compatibility between planned projects and personal interests	2.3	11.0	5.2	5.2	3.5	—
Inappropriate or unpopular allocation of personnel	2.3	1.5	5.0	8.2	3.3	—
Unpopular arrangement of working hours	—	—	3.3	—	—	—
Inappropriate or unclear allocation of equipment	—	—	8.3	—	—	—
Unequal allocation of ancillary staff	—	—	5.0	—	—	—
Imprecise determination of relative contributions to joint work	10.5	6.1	—	—	9.2	10.5
Carelessness and mistakes of subordinates	8.8	4.4	—	—	4.3	—
Incorrect allocation of responsibility and work between researchers	—	—	—	17.0	—	10.5
2. *Connected with characters of researchers*	54.4	45.5	54.1	40.9	54.5	11.4
Incorrect or imprecise understanding of their own or others' responsibilities	10.5	5.2	—	—	8.0	—
Unsociable character of initiator of conflict	7.6	5.0	10.0	14.6	6.8	—
Lack of discipline	7.6	8.0	5.0	9.0	7.0	—
Researcher taking over results of others' work	5.3	11.2	10.0	7.9	7.0	—
Dishonest behaviour	4.7	2.5	10.0	—	6.8	—
Laziness	4.1	2.3	8.3	—	6.0	—
Hostility	3.5	6.2	3.3	—	3.3	—
Tactlessness	11.1	5.1	7.5	9.4	9.5	11.4

SOURCE: A. I. Shcherbakov, *Effektivnost' nauchnoi deiatel'nosti v SSSR. Metodologichesky aspekt.* 'Ekonomika', Moscow, 1982, p.148

The effects of such oppressive bureaucracy on scientific personnel are two-fold.[183] Firstly, particularly among young researchers we find a considerable degree of apathy – not just political apathy, but apathy towards their work as well. For the large proportion who entered science 'by chance' this would probably come naturally. For those who entered science with high hopes of creative success, only to be frustrated by bureaucracy and a promotion ladder choked by the boom in recruitment since the late 1950s,[184] and therefore effectively doomed to working as a laboratory assistant for years to come,[185] the apathy might have come via the dissidence bred by 'over-inflated ideas of one's role in society'.[186] The result is the scene described in the following pseudonymous letter to *Komsomol'skaia pravda*:

> You will not believe this, but our working day proceeds thus: the men talk about football or the cinema or play ping-pong. Sometimes they do some work. The women go shopping, drink coffee, knit, sometimes write something. While I write this letter there are three people in the room besides me. One young man, a candidate of science, is sitting at his desk looking through foreign magazines. He copies out various items on cars, aeroplanes, sport and so on – it's his hobby. He copies it out or clips it and files it away. An 'encyclopaedist'. Two others – not candidates – are writing something. Two are standing on the stairs smoking. One woman, a candidate, has gone off on trade union business. I don't know where the rest are. They wander in and out. . . . They work very little. I would guess about two hours a day. . . . I could change jobs. But in my field there are two other institutions and I've spent some time in both. I didn't like them. Everybody fought between themselves and they all tried to palm their work off onto others. Here we have such a friendly collective. They're very nice people. Sometimes our room is full of laughter.[187]

On the other side, we find an entirely different type of researcher, the wheeler-dealer science manager, the person who adapts to the system and turns it to his own benefit. Because appointment to any managerial position in the Soviet Union requires party approval, science administrators tend to be characterised by complete loyalty to the party. Once they occupy a managerial position they can use that background of party loyalty and the enormous bureaucratic powers that accrue to them to establish themselves as virtual dictators in their area of control. This can mean obtaining a scientific reputation from one's subordinates, for those with little scientific talent or time for research, or an opportunity to protect their work from competition for those with talent. The result – Lysenko-type 'monopolies', even if without the ideological and murderous trimmings – can have only a bad effect.

However, if the case of plate tectonics, described above,[188] can be taken as a guide, we should not exaggerate the extent of these monopolies. For there are factors working to offset the dangers. The party realises that it needs to balance its desire for loyalty with a need for quality research. It has throughout its history accepted as leading scientists people who have not shared its values,[189] and, conversely it appears to expect even from its most loyal managers a degree of competence, at least in management if not in research. More work needs to be done on Soviet science managers, but it appears that they are required to keep up some sort of scientific reputation, in order to maintain their status in the eyes of the party (to say nothing of the international scientific community). Since in many cases they have neither the talent nor the time to earn such a reputation honestly, they rely on the goodwill of the scientific community to accept their 'ghosted' articles and books, and most important of all, to vote in secret ballot for their membership of the Academy. To obtain this cooperation from the scientific community some concessions have to be made – allowing 'unreliable' researchers to continue work in 'unreliable' disciplines, giving scientists greater freedom in their research, and of course looking after them in terms of funding and equipment. One of the best descriptions of the tightrope such science managers walk is found in Raisa Berg's memoirs, in which she recounts the career of a small-town veterinarian turned academician and institute director, D. K. Beliaev. She says of him: 'He always played the role of the behind-the-scenes director of a puppet theatre. His diplomatic capabilities were developed to an improbable degree.' Berg fought with him, and won, over such things as hours of work and attendance at ideological seminars. His election as a corresponding member of the Academy, in July 1964, was seen as a great victory for the barely rehabilitated discipline of genetics, and it was he who gave Berg a copy of Zhores Medvedev's *samizdat* book on Lysenko. However, he refused to resist demands for purges of *podpisanty*, and in the end had Berg dismissed. However, as she says: 'In setting me up in such a way, Beliaev was running a risk. His prestige among the academicians could suffer, his reputation as a fighter for true science would be shaken. And he still awaited election as a full member of the Academy.'[190]

One cannot ignore emigre accounts of unscrupulous institute directors using every form of bureaucratic manipulation to get their own
way and usually to protect simultaneously the narrow interests of the
party.[191] However one is also impressed by the degree to which these
phenomena are offset by two features of modern Soviet science.
Firstly, the party seems unwilling to grant complete power to any
single scientific authority; secondly, the degree to which what we
would call the 'traditional' scientific values of individual creativity
and scientific freedom are retained, both among the rank-and-file
researchers, as reflected in survey data,[192] and among the senior scientists of the Academy, as can be seen in their speeches, writings and
actions. It is worth speculating briefly, on how the scientific community
manages to retain these traditional values. A cynic might say that
there is nothing amazing about an academic realising that having a
job where you can do anything you like is something to be fought
for.[193] The less cynical would say that those 'traditional' values really
are the best for scientific endeavour, and any genuine scientist and
even some politicians will realise that very quickly. But it might be
worth quickly considering the possibility that, whether these values
are in fact best for science or whether they are outdated in the era
of 'big science', the reason that they are held is the continued existence
of a still 'bourgeois' scientific intelligentsia.

I have demonstrated elsewhere that the old 'bourgeois' scientists
were left relatively unscathed by the *vydvyzhenie* movement of the
First Five Year Plan and were certainly not replaced in positions of
leadership by young proletarian party members.[194] They were assimilated into the Soviet political and social system, many of them even
joining the party in the 1940s and 1950s. Although the huge expansion
in the numbers of people employed in scientific research since the
mid-1950s means that many people must have been recruited from
outside the intelligentsia, it is probable that a reasonable proportion
is made up of the children and grandchildren of the old 'bourgeois'
intelligentsia. Surveys show that over 50 per cent of Soviet researchers
are children of members of the intelligentsia,[195] while their children
in turn are most likely to go into intelligentsia jobs.[196] The children
of natural scientists are particularly likely to go into the natural sciences.[197]

It seems probable that the scientific community has been and still
is to some degree a self-reproducing one, and one wonders to what
extent its values are self-reproducing, and to what degree those values
have survived from pre-revolutionary or pre-Stalinist times.[198] Any

tendency of this kind would be reinforced by the traditional importance in Russian science of 'scientific schools' based on a single influential scientist or branch of a discipline,[199] and informal means of communication within the scientific community. Lubrano draws out well
the possible political significance of the latter.[200]

When we mention the 'ideology of science' in the Soviet Union,
most non-specialists probably think first of the ideological control of
scientific knowledge, best exemplified by the declaration of genetics
to be a bourgeois pseudo-science because it talks of random mutations
which are clearly outside the deterministic boundaries of dialectical
materialism.[201] We have seen that in fact ideological control of this
type is probably not the major factor in the Soviet ideology of science.
Certainly the party arrogates to itself the total right to determine
which science is genuine and which is false, a right only slightly reduced
by the ideological innovations of the last couple of decades. However
it is a right that is relatively rarely used, even if with devastating effect
on those rare occasions.

There might be grounds to argue that this right of ideological
control is more than offset by a Bolshevik belief in science as the
source of massive economic and social change, including the eventual
achievement of communism. This has provided an impressive,
although by no means unique willingness to fund research, train personnel and listen to what scientists have to say.

When we come to the values of scientists, particularly 'individualism' versus *'kollektivnost''*, we face the problem of determining
which is in fact the most desirable value for modern science. One
need not be surprised that scientists tend towards 'individualism'. But
this does not necessarily mean that this is what is most needed for
scientific advance. In the era of 'big science' clearly some ability and
desire to work collectively is essential. Nevertheless it is interesting
that most students of Western scientific values, while recognising the
often disruptive effects on R & D projects of scientists' individualism,
still accept such values as legitimate and deserving understanding.
Indeed conscious efforts should be made to fit organisational patterns
around such individualism.[202]

The Soviet authorities, in contrast, make an ideological and organizational virtue of *kollektivnost'*. This has a primary negative effect on
scientists by directly attacking closely held values, and a secondary
negative effect by leading to an overwhelming tendency to over-man-
age. This tendency is reinforced by a traditional official belief, rooted
in the ideological principles of socialist planning, that duplication and

competition cannot be compatible with efficiency and that decentrali-
sation and individual initiative are therefore best avoided. When all
this is combined with an everpresent consciousness among scientists
of tight and degrading political control, it is not surprising that a
serious morale problem follows. The Bolshevik commitment to science
at the macro level has been strong enough that significant scientific
development has taken place despite these serious problems at the
micro level. However, as resources and marginal rates of return decline
in the economy as a whole and in science specifically, the micro level
problems of declining morale and bureaucratic stifling of creativity
are coming to play an ever greater role, despite the impressive ability
to survive of traditional scientific and Bolshevik values.

2　Leading Party Organs and Science Management

This chapter examines in broad rather than specific detail the role of leading party organs in science management, particularly the making of major science policy. Lack of information means that it will be more a study of the institutions involved than a precise description of the process itself. Emphasis will be given to the role played by the Politburo and the Central Committee apparatus. Equal emphasis will be given to the influence of scientists on major decision-making. The findings of the latter examination will play a major role in determining our preference for any particular model of the Soviet political system.

There is no denying that the Politburo is the major policy- and decision-making body in the Soviet Union.[1] The most important policies and decisions affecting science and indeed most areas of Soviet life are set out in Central Committee decrees, often issued jointly with the Council of Ministers.[2] It can be safely assumed that these decrees are at least vetted by the Politburo. They cover major reorganisations of the Academy of Sciences and other science management bodies, the setting out of the latest line on the proper relationship between politics, ideology and science, and the granting of priority to particular branches of science and technology. Sometimes decrees are directed to a specific institute or party organisation, although even in these cases their relevance to the science community as a whole is well understood.[3] They are usually put into effect by more detailed decrees and instructions from the Council of Ministers or agencies such as the State Committee for Science and Technology (GKNT), Gosplan or the State Committee for Labour.

Since the death of Brezhnev we have been able to get a better idea of the concerns of the Politburo from the brief reports of its meetings that have been published in the press. We find that a major decree such as the 1983 joint Central Committee and Council of Ministers decree 'On measures for accelerating scientific–technical progress in

57

the economy', published in *Pravda* on 28 August, was discussed in the Politburo meeting reported on 20 August. We cannot say definitely whether a Politburo discussion always leads to a decree, since decrees are not always published. Thus in 1984 the Politburo was reported as having discussed the wider use of rotor technology in Soviet industry, in what seems to be a case of the Politburo attempting to give priority and authority to an area of technological development which does not have the full support of the R & D community. No decree has been published, yet one has been referred to in the press.[4] Microbiology has been discussed at recent meetings without decrees appearing.[5] At times the Politburo will give instructions to other party bodies to take action on a particular issue. Presumably in these cases an internal party instruction is issued without a formal decree being seen as necessary. In 1982, for example, Shcherbitsky, Ukrainian party first secretary and Politburo member, reported that the Politburo had instructed the Ukrainian party to hold a plenum on progress in implementing goal-oriented research programmes.[6] There have been recent important Central Committee decrees on the Urals Science Centre, the Institute of Economics and on science management in Leningrad,[7] which have been published without reported prior discussion by the Politburo. Nevertheless, one assumes that Politburo members are at least made aware through established formal means of the existence and text of these decrees before publication.

If they are not discussed in the Politburo one assumes that they are at least discussed at Central Committee secretariat level. The Central Committee secretary with most direct responsibility for Academy research at least is Zimianin, who is not a member of the Politburo. Recently the Georgian party started publishing reports of its secretariat meetings. These indicate a high degree of involvement in the day-to-day operations of major enterprises and organisations. It is claimed in Soviet sources that it was at secretariat level that the Ukrainian and Siberian Academies were examined in 1976.[8]

The picture is about what we would expect. The Politburo is a body with an enormous range of responsibilities, and we would expect it to concern itself with only the most important scientific issues. When it does take action, even if only in the form of a Central Committee decree, one can expect a strong and immediate reaction. A. A. Baev, the academician-secretary of the Academy's Biochemistry Department, described the 1974 Central Committee and Council of Ministers decree on molecular biology as one of the most important organisational measures of recent years, which 'as it were removed the brakes from the development of physico-chemical biology and gave it enormous moral force'.[9] The Academy's official historians claim that the February 1975 decree on preparations for the Academy's 250th anniversary celebrations had an important stimulatory effect on the organizors.[10]

However, such a positive reaction cannot be guaranteed. There had already been one decree issued on the celebration of the Academy anniversary, which seemingly had so little effect that the celebrations had had to be abruptly postponed at the last moment.[11] In 1985, in an interview in *Sovetskaia Rossiia* with a leading proponent of rotor technology, details were given of *continuing* inertia in the implementation of the new technology despite the perhaps pointed reference to the September 1984 Politburo discussion of the matter.[12] This is a matter which will receive some more general attention later in our discussion.

Such a bald review of decrees and agendas does not perhaps fully convey the personal interest that Soviet party leaders have always taken in scientific and technological matters. In Chapter 1 reference was made to the traditional Bolshevik 'cult of science'. It is a cult which has been reflected by each leader in his own particular way. Lenin felt himself competent to write on the philosophy of the natural sciences, while he had an amateur's interest, according to Krzhizhanovsky, in such things as X-rays and perpetual motion experiments.[13] As his articles and correspondence show, he was determined to provide the best possible conditions for science and scientists during difficult times.[14] While Stalin was notoriously suspicious of all intellectuals and many scientists suffered at his hands, he was also aware of the importance of technological development and on the whole made careful, if brutal, use of his scientists. D. Anastasyn and I. Voznesensky report a case of Stalin supporting a scientist in a dispute with Beria, and reportedly telling Beria: 'You must listen to specialists.'[15] Arvo Tuominen also writes that Stalin 'had a group of good advisers and he listened to them. I personally had occasion to observe that he had a great respect for technical experts, among others.'[16] Stalin also of course expected scientists to listen to him,[17] and was happy to be called the coryphaeus of science and make important contributions to linguistics.

Judging from his own memoirs and those of others, Khrushchev of all Soviet leaders had the closest personal attraction to science, with his memoirs suggesting that he enjoyed the company of scientists, be they famous (Kurchatov), infamous (Lysenko) or unknown

(Mikhailov).[18] Arnosht Kol'man describes in nice detail the child-like fascination with which Khrushchev regarded scientific gadgets in the 1930s.[19] We have no evidence of Brezhnev's personal attraction to science, but it was nevertheless Brezhnev who presided over the flowering of the scientific–technical revolution, while scientists feature prominently in his memoirs.[20] While Andropov did not have time to exhibit any particular commitment to science as head of the party, he had in his earlier days gathered around him a coterie of social scientists.[21]

It could be said that Chernenko stood out from his predecessors in his lack of apparent connections with science and scientists. Indeed it is possible that Chernenko's earlier assertions of the importance of 'workers' rights' was an indication of an antipathy towards intellectuals. But there is no sign that even if such antipathy did exist that it was allowed to have much effect on the long tradition and the empirical necessity of a commitment to science. So far all we can say about Gorbachev is that he has made the usual positive references to the role of science and technology in Soviet development, and sounds reasonably serious about doing something about it.[22]

It need not only be the General Secretary who has a commitment to science among Politburo members. Among lesser members of the present or recent Politburos it is generally considered that Romanov and Shcherbitsky have a particular interest in science and technology. Romanov presided over many experiments in science management in Leningrad,[23] while the prominent role played by Boris Paton, the president of the Ukrainian Academy of Sciences, in developing and publicising the Ukrainian model of R & D management seems to owe much to the support of Shcherbitsky.

Another aspect of the scientific connections of the top leadership that deserves brief mention is the number of children and relatives of top political leaders who work in the academic world. It appears that academia is one area where influence can be used to gain a prestigious but relaxed job, and is therefore popular with the *synki* (the diminutive form of the Russian word for 'sons', carrying with it something of the pejorative tone of 'little rich kids'). One cannot tell, not surprisingly, from the published evidence whether the *synki* simply occupy sinecures as far from politics as possible or whether their family relationships lead to political influence. Some of the more notable *synki* are Dzhermen Gvishiani, the influential management theorist and deputy chairman of Gosplan, who is the son of an executed Beria-supporting KGB general and son-in-law of Kosygin; Anatoly Gromyko, the son of the former Foreign Minister and director of the

Institute of Africa;[24] and the son of Andropov, who studied at the influential Institute for World Economics and International Relations (*IMEMO*) and then worked for the equally influential Institute for USA and Canada (*INSKAN*) before becoming Ambassador to Greece.[25] If we can believe the clearly prejudiced Leonid Vladimirov, the rocket designer Chelomei deliberately appointed Khrushchev's son to a senior position in his design bureau for the political influence it would bring him.[26]

We can hardly expect to get a true picture of the importance of personal relationships from official Soviet sources, while they occur at too high a level to get more than rumour, often malicious, from emigre and *samizdat* sources. However it seems probable that an ambitious relative could gain considerable influence from such personal relationships – that is the nature of Soviet, if not all, societies. But it seems more likely that many with such relationships are happy to distance themselves from politics – that is why they are working in science. One cannot exclude the possibility that, for example, Chernenko's daughter had an effect on his position on the nature–nurture debate in Soviet genetics,[27] or that Shcherbitsky's *naukoved* brother does not influence his approach to science management.[28] But there is no evidence that such relationships are a significant factor in Soviet politics or science policy-making.

A family relationship is not the only type of personal relationship. It was claimed in *Kommunist* in 1979 that the Politburo maintains permanent links with specialists in all branches of science.[29] We have just mentioned Khrushchev's apparent liking for the company of scientists, and the group of social scientists under Andropov on the 1950s and 1960s. We rely above all on Khrushchev's memoirs to get a feel of the importance of the status of individual scientists in getting access to top politicians. He demonstrates how someone of the importance and prestige of Kurchatov, the head of the Soviet nuclear research programme, could get direct access to the First Secretary, it being Kurchatov who was consulted on Petr Kapitsa's request for funding for a particular project. The funding was refused.[30] Kapitsa, with equal scientific prestige to Kurchatov, could get access, but that access was not enough to overcome Khrushchev's disdain for his 'unpatriotic' ideas on national defence. Kurchatov is also credited with successfully insisting to Khrushchev in 1956 that radiation genetics be allowed to continue despite the protests of Lysenko. (Although, as Thane Gustafson points out, military interest in the genetic effects of radiation was probably also of some relevance.)[31]

Another prestigious scientist whose access to Khrushchev was not enough to bring him success in his endeavours was Andrei Sakharov. Between 1958 and 1962 he made a number of approaches to Khrushchev over the halting of nuclear weapons testing. The first, in 1958, was indirect, with Sakharov persuading Kurchatov to go to Yalta to try to persuade Khrushchev to call off the tests scheduled for November 1958. Presumably at this stage he did not feel that he, still a young scientist isolated from the world and inexperienced in politics, could gain access. But by 1961 Kurchatov was dead and Sakharov had some experience of public politically-oriented campaigns behind him, particularly the education debate of 1958. So this time he went directly to Khrushchev. In fact, according to Sakharov, the forum of his protest on this occasion was a meeting between Khrushchev and nuclear scientists called to discuss the Soviet Union's intention to break the three-year moratorium on testing. That is, in being invited to the meeting, Sakharov was being granted access. It was not until the next year, when he was particularly incensed by what he considered to be an entirely gratuitous and technically useless test scheduled by the Ministry of Medium Machine Building (the cover name for the ministry responsible for nuclear affairs), 'acting basically from bureaucratic interests', that Sakharov *demanded* access. This time, after an unsuccessful appeal to the minister, Sakharov managed to get Khrushchev on the phone while the latter was in Ashkhabad. One assumes that Sakharov would have had to use all his status and prestige to make such a call.[32] On both occasions Sakharov had no success. What is interesting about Khrushchev's response is that he did not simply consider Sakharov's demands politically inappropriate, but that he implied that the fact that the demands were made was also inappropriate. While recognizing Sakharov's good intentions, he wrote: 'However he went too far in thinking that he had the right to decide whether the bomb he had developed could ever be used in the future,' and quoted his reply to Sakharov: 'My responsibilities in the post I hold do not allow me to cancel the tests . . . . As the man responsible for the security of our country, I have no right to do what you're asking.[33] He speaks of himself as the person solely responsible for the political fate of the country, and declares that scientists should not expect to become involved. This makes Kurchatov's apparent influence over Khrushchev all the more impressive.

Another interesting case is the 1942 decision to develop an atomic bomb. It is interesting in that the scientist who pushed the hardest for such a decision was almost certainly unknown to the political leadership, but nevertheless he was forced to make a direct approach because of the opposition or scepticism of his more senior colleagues. With the disappearance from Western publications of uranium fission research in 1940, Georgy Flerov became convinced that secret military programmes must be underway both in Germany and the Allied countries, and that the Soviet Union must therefore begin its own work. Like Sakharov he originally tried to work through the scientific hierarchy, putting the matter before the most senior physicists, such as Ioffe, Kapitsa and Kurchatov, in Kazan at the end of 1941. They were unimpressed – they could not see how the work could be done in the current conditions and felt that there were higher priority projects.[34] Flerov made a few more attempts to have the scientists take the matter up, but eventually in May 1942 gave up and in despair turned directly to the government. He approached Kaftanov, chairman of the wartime committee for the coordination of defence research and the science and education representative of the State Defence Committee, the supreme decision-making body during the war. More recent Soviet sources have it that Flerov still did not get the result he wanted, and that eventually he had to write to Stalin himself.[35] At last Flerov's efforts met with success, undoubtedly helped by the government receiving, presumably through espionage sources, news of British, German and American efforts.[36] After consultation with senior scientists the government decided to proceed with a research and eventually development programme.

Beyond the interesting reluctance of the Soviet scientists other than Flerov to recommend or undertake development of the bomb, two points emerge that are of relevance to this discussion. Firstly, Flerov tried hard to work through the usual channels before in desparation making a direct approach. This makes the internal politics of the scientific community particularly important. Some of the apparent internal issues of the time were a jealousy of nuclear research among other scientists, and of Kurchatov in particular, and poor relations between the Moscow and Leningrad schools of nuclear science. Secondly, when the issue is important enough (in this case, and probably in most others, when the military implications are inescapable), a direct approach by an unknown scientist can have results, despite the scepticism, if not opposition, of his more prestigious colleagues.

From these cases we get a reasonably strong impression of individual scientists of high prestige (and not even of high prestige – for example, Flerov, Khrushchev's Mikhailov, and even Lysenko at the beginning of his career) having direct access to the highest decision-making levels. One would do well to pay close attention to

Khrushchev's negative response to Sakharov's efforts to halt atmospheric testing in order to put the significance of this access in perspective. Nevertheless there is good evidence that certain scientists have been able directly to influence decisions in their area of expertise and sometimes even on borderline issues.

The access which these senior scientists are known to have to top political leaders has two further important consequences. Firstly, they are most likely to gain membership, or be able to recommend others for membership, on the numerous commissions and working groups that play such an important role in Soviet policy-making. One assumes that the party exercises its usual right to vet all appointments in these cases. To have good relations with the top party leaders should virtually guarantee appointment to such bodies.[37] Secondly, scientists with such relations with top party leaders gain enormous power in lower-level science decision-making. The person who is most easily able to take a case to, and gain support for his view from, the top party leadership will be most likely to enjoy deciding authority even over decisions that are made at lower levels. This is best illustrated by the description by the Georgian physicist Elevter Andronikashvili of his representations to Kurchatov to get funding for cosmic ray research. He first approached Kurchatov in 1954 and was turned down. In 1955 he returned with a better prepared and more detailed proposal, worked out jointly with a number of senior scientists, and this time got Kurchatov's approval. That approval, which was announced by Kurchatov at a meeting called by deputy chairman of the Council of Ministers Malyshev, apparently required the transfer of funds from the accelerator programme. We cannot be certain where the decision was made, but Andronikashvili certainly gives the impression that it was made primarily by Kurchatov.[38]

Within his native Georgia Andronikashvili had sufficient confidence in his own contacts that he went to republican Central Committee secretaries himself. In the early 1950s he was dissatisfied with Gosplan's projections of the growth in the number of research workers and complained about it to the Georgian party second secretary, P. V. Kovanov. Kovanov advised him to prepare a submission for the All-Union Central Committee, which he promised to send on.[39] Many years later Andronikashvili had a very old acquaintance, Viktoriia Moiseevna Siradze, as Georgian Central Committee secretary for ideology. He had first met Siradze when she was in her mid-20s and had moved from being a secretary of the Tbilisi *komsomol gorkom* to the first secretaryship of the party *raikom* in the *raion* where

Andronikashvili's institute was situated. When the institute was moved to another *raion*, Siradze moved over to that *raikom* (whether by design or not is not made clear). She was later a deputy chairman for culture of the Georgian Council of Ministers, before becoming Central Committee secretary. At all stages of her career Andronikashvili used her to get support for his proposals.[40] Of course it is no great discovery in Soviet studies to find that personal acquaintance going back over many years maintains its political significance as people move into the highest levels of the system. But it is worth mentioning that the process also works in the case of scientists.

One assumes that access to top leaders is on the whole limited to a small number of scientists with influence over a narrow range of issues. We must therefore devote considerable attention to the more routine and formalised means of scientist input into the policy-making process. We will begin by concentrating on the role of the Central Committee – its elected membership and apparatus – in policy-making.

CENTRAL COMMITTEE MEMBERSHIP

One is uncertain how much attention to give scientist membership of the Central Committee, since one does not want to exaggerate the political significance of that body. Khrushchev's 1957 experience, when a Central Committee vote overturned a Politburo decision to remove him from power, indicates that the Central Committee does have an important potential political role, but there are so few scientists in the Central Committee that it is unlikely that their numbers will ever have any significance, even if a situation of the 1957 type were to arise again. There is presumably some informational role to be played by Central Committee membership – the transmission, often in closed and perhaps informal circumstances, of sensitive information from the political leadership to the nation's elite and conversely the transmission of advice and requests from the elite up.[41] However it seems most probable that the main function of Central Committee membership is as an indicator of status of either an individual or the institution he represents. An analysis of scientist membership is instructive in this regard.[42]

A minor problem that arises in assembling the data is determining who can be classified as a scientist and who not. Breakdowns of Central Committee membership by different authors show different numbers of scientists, depending on the method of classification used. I have tended to work according to position held at the time of election, thus excluding such academicians as Ponomarev, Ilychev and Korneichuk. I have included such dubious philosophers as Egorov and Iovchuk when they have occupied academic posts, but not when they have occupied editorial or diplomatic posts. I have included science managers, including the head of the Central Committee Science Department. The uncertainties of classification make me reluctant to give total scientist membership levels, but they might be useful in providing an idea of general orders of magnitude. It is worth noting that the percentages of Central Committee membership working in science, in recent years anyway, are not all that far below the percentage of party members working in science. In 1976 4.3 per cent of party members worked in science; in 1981 4.4 per cent.[43]

Another noteable feature of the table is the very low level of Central Committee representation in 1956.[44] The 1956 Central Committee seems to mark a watershed between the years when the ideologists were the only academics in the Central Committee, and the later years when natural scientists gained an increasing number of places. It was in this year that Kirillin, the first 'real' scientist, appeared, in the Auditing Commission, although he was there in his capacity as head of the Central Committee Science Department rather

Table 2.1 Scientist membership of the Central Committee, 1951–81

	1951	1956	1961	1966	1971	1976	1981
Member	3 (2.4)	1 (0.8)	3 (1.7)	9 (4.6)	8 (3.3)	11 (3.8)	12 (3.8)
Candidate	1 (0.9)	1 (0.8)	5 (3.2)	2 (1.2)	5 (3.2)	6 (4.3)	4 (2.0) *
Auditing Commission	3 (8.1)	1 (1.6)	—	—	2 (2.5)	3 (3.5)	3 (4.0)
Total	7 (2.5)	3 (0.9)	8 (2.0)	11 (2.5)	15 (3.1)	20 (3.9)	19 (3.3)

* Note that Academician Chazov was moved from candidate to full membership at the May 1982 Central Committee plenum.

Note: Scientist membership as a percentage of total Central Committee membership is given in brackets.

than as a specialist in high-temperature physics. By 1961 the ideologists were being joined in significant numbers by practising researchers and science managers. In this year Rudnev, chairman of the State Committee for the Coordination of Scientific Research (the predecessor of GKNT), and Keldysh, president of the Academy of Sciences, were full members (Fedoseev, the director of the Institute of Philosophy, being the only other 'scientist' full member); while Paton, the director of the Ukrainian Academy's Institute of Electrical Welding, Lavrent'ev, chairman of the Siberian branch of the Academy, Semenov, director of the Institute of Chemical Physics, and Kirillin were candidates (Frantsov, rector of the Academy of Social Sciences, being the other candidate). Since then the scores for researchers and managers versus ideologists and other social scientists (for full, candidate and Auditing Commission members) have been 1966 F 6:3, C 2:0; 1971 F 6:2, C 3:2, A 1:1; 1976 F 9:2, C 3:3, A 3:0; 1981 F 8:4, C 4:0, A 2:1. Thus, if we are looking for signs of the increasing status of genuine science in Soviet decision-making we should look not just at the quantitative changes, but also at the qualitative changes. The latter are even more striking than the former.

There appears to be a limited number of science establishment positions that bring *ex officio* Central Committee membership. Since Keldysh became president of the Academy of Sciences in 1961, he and his successor A. P. Aleksandrov have always had full membership of the Central Committee. It is interesting that Keldysh's predecessor Nesmeianov never held a seat in the Central Committee at any level, despite being the first party member to be elected Academy president. How much that was a factor of the changing status of the Academy or of individual presidents – there has always been some doubt about the acceptability of Nesmeianov to the party – must be left to a political history of the Academy. But we are now almost certainly justified in describing the presidency of the Academy as *ex officio* bringing full Central Committee membership.

To offset the full membership of the Academy president the chairman of GKNT or equivalent body has also had full membership since 1961. Rudnev, Kirillin and Marchuk, the three incumbents of this post, gained full membership only after their appointments to the job. (Khrunichev, the first chairman, had full membership before his appointment, but died very soon after.)

The other member of the *troika*, the head of the Central Committee Science Department, had full membership while the post was occupied by S. P. Trapeznikov, that is, from 1966 to 1983. His predecessor,

Kirillin, had been a member of the Auditing Commission in 1956 and a candidate member in 1961, while Kirillin's predecessor, A. M. Rumiantsev, was already a full member when he took over the Science Department. Trapeznikov's replacement, Vadim Medvedev, is a member of the Auditing Commission, status gained and held while deputy head of the Central Committee's Agitprop Department and rector of the Academy of Social Sciences. It will be interesting to see if he gains full Central Committee status at the next Party Congress.

The chairman of the Siberian branch of the Academy has always had some level of Central Committee membership. Lavrent'ev was made a candidate member in 1961, the first opportunity after his appointment as founding chairman of the branch (as an institute director and Academy academician-secretary previously he had not enjoyed Central Committee membership at any level). He retained his candidate membership until 1971, after which he retired. His place as chairman and candidate Central Committee member was taken by Marchuk. With Marchuk's promotion to the chairmanship of GKNT in 1980, Koptiug was elected chairman of the Siberian branch. In 1981 he was elected to the Auditing Commission. This is perhaps due more to his newness in the chairman's post and relative obscurity than to any downgrading of the Siberian branch in the party's eyes.

Ever since becoming president of the Ukrainian Academy of Sciences Boris Paton has been a full member of the Central Committee (1966–81), having been elected a candidate member in 1961 while director of the Paton Institute of Electrical Welding (named after his father). His full membership seems likely to be a case of *ex officio* membership, but it will take the replacement of Paton, a scientist who seems to be the darling of the party establishment, to confirm this, and particularly whether Paton's full membership versus the candidate or Auditing Commission status of the Siberian chairman reflects the relative standing of the Academies or of their chairmen.

We therefore have a limited number of *ex officio* posts, but most of the major elements of the Soviet research establishment are covered. It should be added that Eliutin, until recently the Minister for Higher and Specialized Technical Education, has full membership, as a representative of *vuz* research. Branch research is less well off, in that as direct representatives it has no more than various ministers, with Soviet branch ministries being notoriously unenthusiastic about research and innovation. There is however the chairman of GKNT as a full member, and one of his deputies, Efremov, as a long-time member of the Auditing Commission (1971–81).

A very important branch of research is defence and space research. It is well represented in the Central Committee. The rocket designer Mikhail Iangel' died in 1971, but at the Party Congress in that year and the previous 1966 Congress he had been elected a candidate member.[45] His colleague in the aviation industry, Petr Grushin, who joined the party in 1931, the same year as Iangel', has been a full member since 1966. Viktor Makeev, like Iangel' and Grushin, is a product of the Moscow Aviation Institute, and he also enjoys Central Committee status (candidate in 1971; full member since 1976). He is described in his laconic Soviet biography as 'a specialist in the field of mechanics, who has developed original areas of research in the field of machines and mechanisms. V. P. Makeev has solved a range of problems connected with hydrodynamics, aerodynamics, strength and other problems relevant to the construction of automated systems.'[46] This, together with his work in enterprises of the Ministry of General Machine Building, suggests work in missile technology, presumably in the missile plant in Cheliabinsk, the region he represents at Party Congresses. Valentin Glushko is a graduate of Leningrad State University and a well-known specialist in gas dynamics who worked with Korolev on jet engines during the Second World War, and now concerns himself with rocket and space research as director of the Gas Dynamics Laboratory. He has been elected a full member of the Central Committee at the last two Party Congresses. I have no information specifically linking V. F. Utkin to defence or space research, but what biographies we have of him sound very similar to those of the defence scientists just mentioned. He is a 1952 graduate of the Leningrad Mechanics Institute, a doctor of technical sciences and a member of the Ukrainian and All-Union Academies of Sciences. He is described as a researcher in the field of mechanics who since 1952, that is, since graduation, has worked as an engineer in machine-building enterprises, presumably in Dnepropetrovsk, since that is the region he represents at Party Congresses. The lack of detail about the enterprises he has worked in suggests that they are defence-oriented. He has been elected a full member of the Central Committee at the last two Party Congresses.

Thus at the last two Congresses four defence scientists have been elected full Central Committee members, of a total of twelve researchers who are full members and eighteen at all levels. Remembering the large numbers of military officers already representing the military establishment in the Central Committee, this is an impressive proportion just for military research. Presumably the four are consultants to the Politburo on defence R&D matters. It is worth noting that they appear to come mainly from the missile technology branch of defence R&D. This could lead us to the unremarkable conclusion that missiles are the dominant branch of the Soviet defence establishment.

The only other scientific group ever to have rivalled the defence scientists in terms of Central Committee membership are the ideologists. At the beginning of this section we saw how in the early 1950s they had a monopoly of Central Committee 'science' membership. Khrushchev shattered that monopoly, but the ideologists have never disappeared from the Central Committee altogether. They are far better known than the defence scientists. Such philosophers as Pospelov, Egorov, Iovchuk and Fedoseev, and economists like Rumiantsev and Ostrovitianov, have been features of the Soviet academic, political and journalistic scene for years.[47]

One small group that is clearly separate from the ideologists despite its members' social science background is the foreign policy group. This group includes only two people who can be classified as scientists, Georgy Arbatov and Nikolai Inozemtsev. Arbatov is the director of *INSKAN*. He eventually made it to full membership in 1981, after being a member of the Auditing Commission in 1971 and a candidate member in 1976. Inozemtsev was director of *IMEMO* until his death in 1982. He also made full membership in 1981, after being a candidate in 1971 and 1976. But while academics, this pair's membership should not be seen as representation of the academic community, but rather of the foreign policy community. A number of non-academic members of this community have also improved their status at recent Congresses, particularly the 1981 Congress.[48] It will be interesting to see how well the foreign policy lobby survives, in terms of Central Committee representation, recent leadership changes.

This brings us to scientist members of the Central Committee whose membership does not appear to be related to any particular post they occupy. It is sometimes claimed that Central Committee members whose membership does not derive from any post they hold, particularly if they are representatives of minor groups such as workers, *kolkhozniki* and perhaps scientists, are simply 'token' members, to show the world that all social groups are represented in the most important Soviet decision-making bodies. In fact in most cases of apparent non-*ex-officio* scientist membership, a pretty clear link can be made to the particular status of the scientist or the special importance of the discipline represented. This applies to most of the seven scientist Central Committee members not yet accounted for. They are N. N. Semenov, a candidate member in 1961; Iu. A. Ovchinnikov, Auditing Commission in 1976 and candidate in 1981; A. P. Aleksandrov, full member in 1966 and 1971 (before his election as president of the Academy of Sciences and consequent *ex officio* membership); B. E. Paton, candidate in 1961 (similarly before his election to the presidency of the Ukrainian Academy); E. K. Fedorov, a candidate member in 1976 and 1981; A. M. Kunaev, candidate in 1976 and 1981; and E. Chazov, candidate in 1981 and elected to full membership at the May 1982 plenum.

Semenov, the pioneer in research into chemical chain reactions and the Soviet Union's first Nobel Laureate (1956), was in 1961, as he had been since its foundation in 1931, the director of the Academy's Institute of Chemical Physics. He was also at the time the academician-secretary of the Academy's Department of Chemical Sciences. This one appearance in the Central Committee would seem to be related to Khrushchev's huge drive at the time for the development of the Soviet chemical industry.[49]

Ovchinnikov's Central Committee status would seem to be relatively easily explained, in that Ovchinnikov, a vice-president of the Academy of Sciences, director of its Institute of Bioorganic Chemistry and chairman of its Section for the Chemico-Technical and Biological Sciences, is the leader of biological research in the Soviet Union, particularly molecular biology. In recent years molecular biology has become as fashionable in the Soviet Union as it is in the West. It seems to have taken over the mantle worn by chemistry in the 1960s, and Ovchinnikov's place in the Central Committee is presumably connected with this. This would be particularly so if Ovchinnikov is devoting as much of his microbiological research to military uses as some Western observers have suggested.[50]

Aleksandrov's election as a full member in 1966, while his official post was director of the Kurchatov Institute of Atomic Energy, is less clear-cut. He was appointed to the post in 1960, but having joined the party only in 1962 he clearly could not have had Central Committee status in 1961. Nevertheless the directorship of the institute in itself seems unlikely to entail Central Committee membership. It is more likely that Aleksandrov's membership was related to his more informal position as head of the Soviet Union's energy research programme.

In 1961, when elected a candidate member of the Central Committee, Boris Paton was not yet the president of the Ukrainian Academy of Sciences and the highly publicized innovator in research management that he is today. However he was director of the Paton Institute of Electrical Welding and the acknowledged leader of Soviet welding research. But this on its own does not seem sufficient reason for Central Committee status. While Aleksandrov's 1966 status, even if not connected with a particular campaign, is primarily the consequence of his role in the Soviet energy campaign, Paton's case seems to be one where the personal status of the individual scientist is of prime importance. His formal and informal status, as the representative of a particular institution or branch of science, was not that of Aleksandrov, but he was presumably already seen as a young, up-and-coming science manager (15 years younger than Aleksandrov, but 10 years longer in the party) who could already in 1961 contribute to science policy-making.

This raises the last point that I want to make in this section. On the whole, analysis of scientist members of the Central Committee shows that membership can usually be tied very directly to a particular post or to representation of a particularly important branch of science. Nevertheless the individual cannot be ignored. While official position might be dominant (certainly if the ideologists are excluded), nevertheless variations according to individuals can be found.

Paton, and particularly his 1961 candidate membership, would seem to constitute one of the cases where the status, prestige and prospects of the individual scientist played as great a role, if not a greater one, than official position or group representation. The other clear case is that of Koptiug, the present chairman of the Siberian branch of the Academy of Sciences and the first holder of that post not to gain candidate status in the Central Committee. His status and prestige were nowhere near those of his predecessors, and consequently he has had to make do with Auditing Commission status. It should also be mentioned that Keldysh retained full membership in 1976, although he had retired as president of the Academy in 1975. Presumably it was felt that he still had something to offer as a senior member of the research establishment with special expertise in rocket and space research.[51]

Two cases where personal contact, rather than prestige alone, seems to have played a role are Kunaev and Chazov. In 1976 and 1981 A. M. Kunaev, president of the Kazakh Academy of Sciences, was elected a candidate member of the Central Committee. None of his predecessors had been members, and he had only joined the party in 1971. The Kazakh Academy is not obviously the second most important republican Academy after the Ukrainian. By some indicators the

Uzbek and particularly the Georgian could be seen as more important. One wonders therefore whether the fact that Kunaev is the half-brother of Dinmukhamed Kunaev, the first secretary of the Kazakh party and a Politburo member, gave him the advantage over his colleagues in other republics.

Academician Chazov, elected to full membership at the May 1982 plenum after gaining candidate status at the 1981 Congress, is known as a brilliant surgeon and medical researcher. He has been a deputy Minister of Health since 1967 and director of the All-Union Cardiological Centre of the Academy of Medical Sciences since 1976. But in the last years of Brezhnev's life Chazov became best known as the late General Secretary's personal doctor. It seems more than likely that it was this qualification that gained him Central Committee status.

One scientist who by my definition should perhaps have been excluded from the count is E. K. Fedorov. He was elected a candidate member in 1976 and 1981. Fedorov first made a name for himself as an Arctic explorer in the 1930s, his specialisation being meteorology. Having joined the party in 1938, he became head of the Soviet meteorological service in 1939. From 1947 to 1955 he worked in the Academy's Institute of Applied Geophysics, becoming director in 1955. In 1959 he became acting Chief Scientific Secretary of the Academy of Sciences, and the following year, upon election as a full member of the Academy, was confirmed in the post.[52] He had a brief and seemingly unhappy term as Chief Scientific Secretary, resigning in 1962.[53] He then went back to his old job at the meteorological service. In 1974 he went from there back to the directorship of the Institute of Applied Geophysics. It was at the next Party Congress, in 1976, that he was elected a candidate Central Committee member. He retained this status in 1981, despite the fact that Iu. A. Izrael' was only a member of the Auditing Commission. Izrael' was Fedorov's successor at the meteorological service, which in 1978 was given greater status by being renamed a State Committee, with Fedorov's institute in fact being subordinate to it. There is nothing about Fedorov's career to suggest great brilliance or its special importance to the Soviet regime, particularly in its later stages. Reappointment to the directorship of the Institute of Applied Geophysics hardly seems sufficient reason for Central Committee status.

However Fedorov did have another entirely unscientific side to his career. He was long active in the Soviet peace movement, becoming chairman of the Soviet Committee for the Defence of Peace in 1979. This cannot explain his Central Committee status in 1976. However he was a deputy chairman of the Committee previously, and in one source is described as first deputy chairman in 1976.[54] It seems probable that his duty chairmanship was enough to give him Central Committee status, particularly when the chairman at the time, N. S. Tikhonov, was not a party member. Thus Fedorov's Central Committee membership was unconnected with his not particularly illustrious scientific career.

Despite the signs of the importance of personal prestige and contacts, the overwhelming impression to be gained from an analysis of the scientist membership of the Central Committee is that membership can be tied relatively directly to official post, group representation or special policy campaigns. We see the growing status of the natural sciences since the death of Stalin, the seeming equal party status of the Academy of Sciences, GKNT and Central Committee Science Department, and the great status of defence R&D.

However it needs to be repeated that there is no reason to believe that Central Committee membership in itself provides any access to policy-making. It is more likely to reflect already existing access, although the prestige that goes with Central Committee membership presumably reinforces the right to that access.

THE CENTRAL COMMITTEE APPARATUS

What form then does access take, beyond the direct personal relationships with top political leaders already described? If decisions are to be made, or even simply approved, at Politburo level, one would expect the Central Committee apparatus to play an important role.

There are two sides to the Central Committee apparatus. It works, firstly, as a giant staff office serving the Politburo and Secretariat. As such it has two essential functions. It controls access to the executive organs it is serving, but at the same time ensuring that those with genuine business get the access required. It also prepares documents for the Politburo and Secretariat – summaries of submissions and memoranda, position papers, and draft decisions and 'directive documents'.

But it has, in addition, an important control function. Like all party organs it has major personnel powers, summed up in its *nomenklatura*, and is also charged with ensuring the fulfilment of policies set by the Politburo. It is expected to exercise these control functions from both a political and an economic or technical standpoint.

The apparatus is headed usually by ten secretaries, making up the Secretariat. As of May 1985 six are members of the Politburo (including two candidates). Each secretary is responsible for a broad area of the system's operations - ideology, agriculture, heavy industry, etc. Each of these broad areas is served by a number of departments. The exact number and functions of these departments are either rather fluid or something on which we do not have enough information to form a picture with complete confidence. However it would appear that there are approximately twenty departments, including such functional departments as the General Department (responsible for party documentation, including routine paperwork for the Politburo), the Organizational Party Work Department (responsible for the supervision of the whole party apparatus and various social organisations), the Propaganda Department, the International Department, the Department for Science and Education, the new Economic Department, and a range of branch departments. These include the Departments for Agriculture and Food Industry, Heavy Industry, Light Industry and Consumer Goods, Machine Construction, Defence Industry, Chemical Industry, Fuel Industry, and Construction and Transport. The total staff of the apparatus, excluding secretarial and ancillary workers, is apparently about 1500.[55]

One might expect the most important department for our purposes to be the Science Department. It was first set up as an independent unit in May 1935, with the breakup of the old Culture and Propaganda Department.[56] The relevant Central Committee decree states that 'Culture and Propaganda [is] to be reorganized, with five departments being set up in the Central Committee apparatus in place of Culture and Propaganda.' The five departments were Party Propaganda and Agitation, the Press and Publishing, Schools, Cultural-enlightenment Work, and Science.[57] The new departments were set up 'in place of' (*sozdav vmesto*) Culture and Propaganda, with Propaganda and Agitation listed as a separate department, suggesting the complete independence of the Science Department. Its full name was the Department of Science and Scientific–Technical Inventions and Discoveries, and it appears to have been responsible for all R&D organisations. It was abolished in 1939 with the restructuring of the Central Committee apparatus along functional lines.[58] Sometime during the war the department is said to have been reestablished, apparently within the Agitation and Propaganda Administration, and then as a sector in the Agitation and Propaganda Department when the whole apparatus was reorganised in 1948.[59] In 1950 the Agitprop Department was split into four, with one of the new departments being the Department

for Scientific and Higher Educational Institutions.[60] Again the fact
that another of the four new departments was the Department for
Propaganda and Agitation suggests the independence of the Science
Department. Sometime around Stalin's death it was renamed the
Department of Science and Culture,[61] and then with Khrushchev's
bifurcation of the Central Committee apparatus into RSFSR and
union–republic hierarchies in 1956 it was put into the RSFSR side,
still independent of the Agitprop Department, as the Department of
Science, Schools and Culture.[62] In 1965, with the dismantling of
Khrushchev's structure, it became the Department of Science and
Educational Institutions.

The department is formally charged with the supervision of the
Academy of Sciences and its institutes, as well as GKNT.[63] Major
Academy events are attended by the head of the Science Department
or more senior party ideologists.[64] The fact that these senior party
officials are usually ideologists provides an indication of the most
interesting feature of the Science Department, its ideological character. Indeed, over the last twenty years it has gained an unsavoury
reputation as a stronghold for conservative if not reactionary forces.
Its head from 1965 to 1983, Sergei Trapeznikov, was well-known as
one of the most consistent neo-Stalinists in the Soviet Union. He is
described in his biography as having been brought up in an orphanage
and having begun work in 1925 as an agricultural labourer hiring
himself out to kulaks. He joined the party in 1931 and began a career
in *komsomol* and party work. One of his first party appointments
was as head of the Culture and Propaganda Department of a *raikom*,
and from then on he was involved in ideological work. From 1948 to
1956 he was director of the Moldavian Higher Party School, and in
1960, after a stint in the Central Committee apparatus, became pro-
rector of the All-Union Higher Party School. His academic discipline
was described as the history of the USSR; he specialised in the history
of Soviet agricultural policy, with such publications to his name as
*The Agricultural Question and Leninist Agricultural Programmes in
Three Russian Revolutions, The Historical Experience of the CPSU
in Implementing the Leninist Cooperative Plan* and *Lenin and the
Agrarian-Peasant Question.* He also put his name to more general
works, with *The Turning Points of History (Lessons of the struggle for
scientific socialism against revisionist tendencies)* being the most indi-
cative of his political stance.

Some of his neo-Stalinist adventures included his 1966 efforts to
get the works of Stalin and a positive evaluation of his historical role

included in university social science courses,[65] his complaints to Brezhnev about the text of the speech prepared for the General Secretary on the centenary of the birth of Lenin,[66] and his displeasure with the volumes of the official *Istoriia KPSS* published with references to Stalin's repressive policies.[67] He is credited with having played a key role in the purges of the Institute of Economics and *IKSI* in the early 1970s, while he combined duty with the settling of personal scores in the harrassment of the 'liberal' party committee of the Institute of History in the mid- 1960s. (The secretary of the committee, V. P. Danilov, was an historian of collectivisation who was seen by Trapeznikov as an academic rival.)[68] In 1968 Andrei Sakharov was sufficiently concerned about Trapeznikov to single him out as 'one of the most influential representatives of neo-Stalinism at the present time' in his essay on peaceful coexistence and intellectual freedom.[69] His is one of the most famous cases of failure to be elected a full member of the Academy of Sciences.[70] His retirement was announced in *Pravda* on 20 August 1983, and despite his advanced age (he was 71 and died soon after)[71] and the dignified manner of his removal, there seems little doubt that it was connected with Andropov's purge of the Brezhnev ideological apparatus.[72]

The new head of the department, Vadim Medvedev, was originally an economist, graduating from Leningrad University in 1951 and working on the staff there until 1956. He then worked as a *dotsent* in the Leningrad Institute of Rail Transport Engineers from 1956 to 1961, and was then faculty (*kafedra*) head in the Leningrad Technological Institute up to 1968. For the next two years he was a secretary of the Leningrad *gorkom*, before spending nine years as deputy head of the Central Committee's Propaganda Department. In 1979 he went to the Academy of Social Sciences as rector, and stayed there until his appointment to head the Science Department. Despite his ideological background, there are signs that he is a more moderate character than Trapeznikov. For example, he tried hard to find some common ground for the two sides in the bitter 1983 debate in the presidium of the Academy of Sciences between the mathematical and political economists.[73] He was elected a corresponding member of the Academy in December 1984 without apparent difficulty.

Of the twelve deputy heads of the department identified since 1960, we have biographical data on eight. Three are clearly ideologists. Evgeny Chekharin, deputy head in the late 1970s, is described as a specialist in the theory of state and law and scientific communism. His previous posts included first deputy head of the ideological department of the (then bifurcated) Central Committee apparatus and rector of the Higher Party School. One of his claims to fame is that he accompanied Khrushchev on his tour of the Manege Art Exhibition in 1962. In 1976 he was elected a corresponding member of the Academy of Sciences. He was appointed deputy Minister of Culture in 1978, and was transferred to 'other work' in 1983. He has written on scientific matters, although as a philosopher. See, for example, his 1965 work *Rol' nauki v sozdanii materialno-tekhnicheskoi bazy kommunizma*.[74] Another former deputy head of the department, Dmitry Kukin (1956–60), was a corresponding member of the Academy, elected in 1964 for his work on the history of the CPSU. He later became deputy director of the Institute of Marxism–Leninism. He died in 1983.[75] Rudol'f Ianovsky was appointed deputy head of the department in 1979, having previously been first secretary of the Sovetsky *raikom* in Novosibirsk (the *raion* in which Akademgorodok is situated) and head of the Science Department of the Novosibirsk *obkom*. He is a doctor of philosophical sciences and a prolific author of articles and monographs on ideological matters, in which he tends to take a hard-line stance.[76] In 1983 he replaced Medvedev as rector of the Academy of Social Sciences.

None of the remaining deputy heads on whom we have information are obviously connected with the natural sciences. Igor' Makharov would appear to have a non-ideological background. He has a doctorate in technical sciences, and in 1974, apparently while still in the Central Committee apparatus, was elected a corresponding member of the Academy of Sciences by the Academy's Department of Mechanics and Control Processes. But the next year he was appointed deputy Minister of Higher Education. This is enough to suggest that his responsibilities in the Science Department were the supervision of education, probably technical education, rather than science. S. G. Shcherbakov, deputy head and first deputy head in the 1960s and 1970s, was recently appointed Minister of Education, also suggesting some education specialisation in the Central Committee, although he often attended Academy and other meetings on science and technology.[77] Oleg Shchepin and Pavel Shirinsky both have medical backgrounds and are recent deputy heads of the Science Department. Shchepin is a doctor of medical sciences and a member of the editorial board of *Sovetskoe zdravookhranenie*, while Shirinsky was head of the medical sector of the department before becoming deputy head of the whole department. As deputy head he attended the 1982 Congress of Medical Workers.[78] Gennady Strizhov is a somewhat atypical case. In 1976 he attended the 25th Party Congress as first secretary of the Kalinin *raikom* in Moscow. Since 1977 he has been a deputy head of the Science Department, although in 1980 he was identified as a secretary of the Moscow City Council of Trade Unions.

We have about the same amount of information at the sector head level. All the sectors I have been able to identify are connected either with education or the social sciences.[79] There are a number of unidentified sectors, but unfortunately we have information on the heads of only four of them. One, Nikolai Kolesnikov, is listed as being a member of the collegium of the Ministry of Higher Education in 1973, six years after being identified as sector head in the Department of Science. In the 1950s he had been head of the student section of the All-Union *komsomol*. It seems reasonable therefore to assume that his responsibilities in the Central Committee were in the sphere of education. Another, Semen Khromov, is a doctor of historical sciences who left the Central Committee apparatus in 1980 to become director of the Institute of History of the USSR.

Interestingly, the other two both show evidence of working outside the social sciences and ideology. One, F. I. Dolgikh, was a sector head in 1968, and then appears in 1972 as head of the Chief Architecture Administration of the Council of Ministers. This hardly makes him a natural scientist, but it does indicate that the Science Department has concerns outside the social sciences.[80]

The only information we have on the other, Anatoly Ermakov, is that he is a doctor of chemical sciences. But this alone is enough to suggest that a sector might exist in the Science Department concerned with chemical science, presumably the chemical research of the Academy of Sciences. According to Semen Reznik in the late 1960s there were two people in the Science Department who dealt with biology.[81] In terms of personnel this is the only evidence we have of a natural science presence in the Science Department, although we do know that the department makes use of academic consultants, including in the natural sciences. Nikolai Dubinin was apparently used as a consultant (*konsul'tant*) on biology until his 'retirement' in 1981.[82]

Despite this the evidence is overwhelming that the Science Department is an ideological organisation, with little expertise in the natural or technical sciences.[83] Its essentially ideological nature is confirmed by an analysis of the heads of such departments at the union-republic level. Using the volumes prepared by Ogareff and Hodnett we see that the previous and later posts of Science Department heads are almost invariably in the ideological and educational spheres (the most popular being Central Committee secretary for propaganda, Minister of Culture and deputy chairman of the Council of Ministers for cultural affairs).[84]

If Academy research is supervised by the Science Department, there seems little doubt that the research institutes of the branch ministries are supervised not by the Science Department, but rather by the same Central Committee department that supervises their ministry. One cannot ignore the Soviety penchant for bureaucratic dual subordination; nevertheless bureaucratic logic would suggest branch department supervision of branch institutes. This seems to be the pattern that holds at regional level.[85] Although referring to a period presumably in the early 1950s, Beliaev writes that the 'direct coordination of the activity of ministries, agencies and scientific institutions in the development of the newest branches of science and technology was carried out by the departments of the CC VKP(b).'[86] The plural 'departments' suggests the involvement of the Central Committee's branch departments. Certainly branch department officials attend conferences and plenums on branch research,[87] while the branch departments supervise goal-oriented research programmes.[88]

Unfortunately we have next to no information on the possible scientific responsibilities or backgrounds of Central Committee branch department personnel. Some emigre evidence suggests that the branch departments have on their staff people with a solid R & D background who both understand and are sympathetic to the needs of applied research.[89] However it should be said that the heads of the departments have typical mixed party–production backgrounds with no hint of R&D training or experience. For example, the heads of the Heavy Industry (V. I. Dolgikh, until recently), Construction (I. N. Dmitriev) and Machine Building (V. S. Frolov) Departments graduated from industrial institutes in respectively 1949, 1948 and 1941, and worked in production enterprises in 1949–69, 1948–64 and 1941–51. They then all had stints as regional party secretaries before moving to the Central Committee apparatus.

Keeping in mind this rather scanty evidence on structure and personnel, what can we say about the functioning of the Central Committee apparatus?

Controlling input into the policy-making process

In the last few years we have been given some details in official Soviet sources on the way that top-level party decisions are prepared and made, details that can be supplemented by emigre sources. For matters that are to be decided by the Politburo brief position papers for each member are required, as well as a draft decision.[90] Such requirements apply to the regular operational decisions made by the Politburo, which are not usually published; they also apply to the so-called 'directive documents', which are nominally issued by the Central Committee and/or Council of Ministers, or by Party Congresses.[91] It is the task of the various departments of the Central Committee apparatus to prepare these documents. They clearly make considerable use of outside help. The most detail we have comes not from the Soviet Union but from Czechoslovakia, but the experience would seem to be transferable. Zladek Mlynar was a secretary of the Czechoslovak Central Committee during the Prague Spring, but in his memoirs describes in some detail how earlier a group of 'liberal' intellectuals helped devise the political and social policies of the Dubcek government. Mlynar had been an academic who also wrote ideological articles for the party newspaper *Rude Pravo*.[92] As a result of these articles he began to be appointed to 'working groups' preparing party documents and position papers. As he says: 'Officially the working groups were the expression of a trend in the party that demanded a higher degree of scientific management in society; in actuality, they often reflected the inability of the apparatus itself to carry out any form of analysis or generalization whatsoever.'[93]

The Central Committee apparatus would order documents from outside, and then put what was received into a form acceptable to the leadership. Frequently this would entail making summaries of the mass of material, summaries which would often in fact be made by the outsider who had prepared the original material. Mlynar describes how he would slip his own ideas inconspicuously into draft resolutions and speeches. If they got through he could then cite or interpret them in such a way as to make further reform proposals possible.[94]

The evidence from Soviet emigre sources suggests that things work the same way in the Soviet Union. Claims are made that scientists, usually social scientists, are closely involved in the preparation of major documents, including the draft Constitution, Party Congress resolutions and major decrees.[95] For an ambitious academic such involvement is clearly a source of considerable status and potential power. Belitsky hints at the same use being made of the power of *TsEMI* to contribute to 'directive documents' as Mlynar described: 'This is not only an extremely prestigious activity. It also gives to the contributor the chance to include in the draft opaque phrases which then become an integral part of a document which cannot be criticized.'[96] Zemtsov refers to the importance A. M. Rumiantsev, the director of *IKSI*, attached to being the director of an institute that had the right to contribute to 'directive documents', if he were to advance his ambition to become Central Committee secretary for ideology.[97] Given that having a special government decree issued is often the only way of getting anything done, to have a Central Committee decree published on one's own institute could be more useful than is at first apparent. While the recent Central Committee decree on the Urals Science Centre was rather critical, the head of the Centre is able to use it to demand higher priority for it, particularly in solving his never-ending construction problems.[98]

The problem faced by scientists who want to influence the policy-making process in the USSR has been the extreme conservatism of the Science Department under Trapeznikov. Its control of the major policy-making institutes of the Academy of Sciences has created problems for those who have wanted to contribute something new to Soviet policy-making. Trapeznikov and Pilipenko, the head of his Philosophy Sector,[99] were able to engineer the downfall of one of the most influential of the 'new' social scientists, A. M. Rumiantsev. Despite his posts as vice-president of the Academy of Sciences, head of its Social Sciences Sector and director of *IKSI*, his full membership of the Central Committee, and reputed backing from influential members of the Politburo, once Trapeznikov and Pilipenko found themselves in a position to use the incriminating evidence they had accumulated on Rumiantsev and his colleagues, he found himself dismissed from all his posts, *IKSI* virtually disbanded, and sociology seriously compromised as a source of ideas on social organisation.[100] Pavel Skipetrov, head of the Economics Sector, with the support of the traditional economists was also able to purge the Institute of Economics of market economists, although he was not able to engineer the downfall of Fedorenko, the powerful director of *TsEMI*. Indeed the latter was able to arrange the transfer of Skipetrov to another post, as head of the Science Department's Universities Sector.[101]

Nevertheless the Science Department was obviously proving a bugbear for social scientists trying to get access to policy-making. One solution has been to work through other Central Committee departments. Following the death of Brezhnev, much attention was devoted to the Department of Relations with Communist and Workers' Parties of Socialist Countries, headed by Andropov between 1957 and 1967. He built up in the department a group of consultants who to this day wield considerable influence in social policy-making circles, whether from academic or apparatus posts. These include Fedor Burlatsky, the strongest proponent of the discipline of political science in the USSR and a former deputy to Rumiantsev at *IKSI*, now head of the philosophy department of the Academy of Social Sciences, a researcher at *INSKAN* and a political commentator for *Literaturnaia gazeta*; and Georgy Shakhnazarov, still deputy head of the Department of Socialist Countries. A. M. Rumiantsev was a member of the group in his capacity from 1958–1964 as editor of the international communist movement's journal *Problemy sotsializma i mira*, while Georgy Arbatov, now director of *INSKAN* but then working in *IMEMO* and later the Central Committee apparatus, was also a member. Other members of the department at the time included Oleg Bogomolov, now director of the Institute for the Economics of the World Socialist System, and A. E. Bovin, the *Izvestiia* foreign affairs commentator. While it might be an exaggeration to describe these people as 'liberal' or 'progressive', they approach the political and social system with a flexibility and desire to make it work more efficiently that has earned them the distrust of conservative elements, including Science Department consultants such as V. S. Shevtsov.[102]

Another group with somewhat overlapping membership is centred in the Central Committee's International Department and the two foreign policy institutes *INSKAN* and *IMEMO*. Although D. K. Simes makes the point that journalists seem to find it easier than academics to push themselves into the foreign policy sphere, this is not to deny the influence of such academics as Arbatov and the late Nikolai Inozemtsev. Other academics such as N. A. Koval'sky, deputy director of the Institute of the International Workers' Movement, and S. M. Menshchikov, deputy director of Inozemtsev's *IMEMO*, now occupy staff positions in the International Department.[103] Jerry Hough stresses the importance for the links between the International Department and scholars of their geographical proximity in Moscow, the similarities in age of officials and scholars, the fact that they were educated in the same institutions, and the participation of Central Committee officials in academic life – writing books and articles, attending conferences, etc.[104] These links presumably make it easier for *INSKAN–IMEMO* staff members to get top-level access regardless of any possible obstructionism from the Science Department, which has formal responsibility for these two Academy institutes. As far as we can tell the *INSKAN–IMEMO* axis has survived the deaths of Brezhnev and Inozemtsev, although how it fares under the foreign affairs dominance of Gromyko and/or Gorbachev remains to be seen.

At one time another important source of access for social scientists to party policy-making was the Central Committee's Information Sector. It was originally set up in 1956 within the Science Department, but in 1966 was put under dual subordination, with the Propaganda Department also being made responsible for it. The sector was specifically set up to provide the Central Committee with an additional source of information to that simply flowing up through the regional party apparatus. The 1966 subordination to the Propaganda Department was presumably to ensure that the information gained was not simply filed away by the then new head of the Science Department, Trapeznikov. The sector was staffed primarily by sociologists, and presumably had close links with Rumiantsev and *IKSI*. Therefore it is not surprising that with the fall from power of Rumiantsev in 1971, the Information Sector lost most of its influence.[105]

Clearly the influence of social scientists varies according to political circumstances. However, there appears to be no doubt that there are close, one might say organic, links between scholars and Central Committee officals. The officials involve themselves in scholarly activities (with books and articles most likely written by the scholars), while the scholars probably provide input into the policy documents prepared within the Central Committee apparatus. There is a limited degree of movement of individuals backwards and forwards between the two worlds. However, little of this interaction has involved the Science Department. Perhaps inevitably because of its control functions, more likely because of the nature of its former head and his subordinates, the Science Department has played primarily an obstructive role in policy-making. This entailed links with some of the most conservative social scientists, but did not require close intellectual interaction with the scholarly community – few ideas were involved. Therefore social scientists looked to other Central Committee departments to interact with and use as a channel to the top: the Department of Socialist Countries for political theorists; the Information Sector for sociologists; and the International Department for foreign policy specialists. It will be interesting to see if the Science Department's position changes under its new head.

It is impossible to say what influence these people have actually had over policies that have been adopted and implemented. We have certainly not witnessed a social or political revolution in the last twenty years. But the social scientists of whom we have been speaking have not been advocating a social or political revolution. They have demanded no more than a serious and objective collection and analysis of information, a greater realization of the feedback mechanisms operating in a complex system, and a concept of planning that goes beyond crude production output targets. These ideas have at the very least moved from the academic world into the vocabulary of political discourse in the Soviet Union, and probably have had some marginal influence on the way the system operates.[106]

These cases have all been concerned with the social sciences. When we turn to the natural or technical sciences we are struck by the total dearth of information. This is perhaps not surprising given the apparent lack of scientific expertise in the Central Committee apparatus, particularly the Science Department. One assumes that the documents going to the Politburo pass through the apparatus,[107] but the indications are that serious consideration of policy options takes place either in direct contacts between senior scientists and top party leaders or within the state apparatus.

In general the findings for both the social and natural sciences support those of Peter Solomon on criminology. He found, somewhat to his surprise, that the Central Committee apparatus does not play as big a policy-making role as the relevant state organisations. It does not initiate proposals leading to decisions and does not even perform the basic staff work. Ministerial officials keep Central Committee officials informed of their thoughts and preparatory work on new policies and seek their opinions. But according to Solomon's informants the apparatus' duties are 'mainly supervisory and concerned with questions of cadres and policy execution'.[108]

Control functions – personnel

Cadres have long 'been everything' in the Soviet Union. The party's obsession with personnel goes back to the early days of the regime, and has now reached the stage that the word *nomenklatura* is virtually synonymous with the Soviet political system.[109] The *nomenklatura* is of course a list of posts maintained by each party organ. Appointments to these posts are made or vetted by the party organ. There is also a list of individuals considered suitable for appointment to the listed posts. As befits what is seen as a fundamentally important feature of an obsessively secretive society, we have little information on even the posts listed on *nomenklatury*, much less on how precisely the system works. But one is inclined to agree with Zhores Medvedev that with the decline in the interference of the party in the content of science, its personnel powers have gained greater relative importance. Medvedev notes that although the party no longer interferes in research itself, 'the party and state apparatus is more than ever involved in deciding *who* should take part in particular scientific and research missions.'[110]

One guesses that the top science management positions – president of the Academy of Sciences and chairman of GKNT – would be disposed of by the Politburo. Lower-level executives of the Academy – its academician-secretaries and institute directors – are quite possibly on the *nomenklatura* of the Central Committee secretariat. Lists of posts on Central Committee department *nomenklatury* include deputy directors and middle-level management posts (department, sector and laboratory heads).[111] This probably applies only to major central institutes. In sensitive social science institutes such as *TsEMI* and *IKSI* all managerial posts are probably dealt with at Science Department level.[112] Directors of the more important branch institutes would be on the *nomenklatura* of the secretariat or the relevant branch Central Committee department.

We need to pay some necessarily highly speculative attention to how exactly the *nomenklatura* system works. It seems unlikely that, particularly in the natural sciences, appointments are made on the initiative of Central Committee apparatus workers, on the basis of casual impressions picked up more or less at random. It is possible that the apparatus maintains a list of politically suitable candidates for possible appointment to, say, the directorship of a particular institute. But it also appears probable that the apparatus is presented with a recommendation, or perhaps a shortlist, from within the scientific hierarchy. The apparatus would make the admittedly important political check and take account of all the pressures of bureaucratic politics. But one doubts that it would be in a position to make scientific judgements on particular candidates. Even many of the 'political' factors would have been taken into consideration before the matter got to the apparatus. One indication that this is indeed the procedure, although admittedly a very old one, comes from the minutes of a 1931 meeting of the directorate of the Leningrad Physico-Technical Institute, at which the setting up of a Urals filial of the institute was being discussed. The meeting 'directed A. A. Tikhomirov to prepare the matter of the placement of a candidate for the post of director of the Urals FTI, confirming (*soglasovav*) the question with VSNKh and the party Central Committee.'[113] In the social sciences, where the apparatus workers appear to a far greater degree to be linked to the scholarly community themselves, they would have greater incentive and capacity to engage in the ideological–factional disputes of the community and to use more initiative in personnel appointments.

Two more speculative comments can be made about party control of personnel movements. Firstly, what is the relative strength of the major science bodies, particularly the Academy, versus the Central Committee apparatus in the case of conflict over a particular appointment? The prestige of the top scientists, plus their access to political leaders, leads one to suspect that the scientists might have the upper hand, and that within reason the Academy could insist on a particular appointment. The rise and fall of *IKSI* showed, unsurprisingly and conclusively, that academic purges cannot take place without approval from the top, regardless of the attitudes of the Central Committee apparatus.

Further, in a highly specialised scientific world the range of possible candidates might be narrow. One should not doubt the willingness of Soviet leaders seemingly to spite themselves for reasons of ideological or political control. But if the leadership accepts the need for quality research in particular areas – something which they seem increasingly forced to do – some personnel appointments are virtually made for them. We have already seen that even the most party-beholden science managers feel themselves obliged to take into account to some degree the values and interests of their scientific colleagues. The arrest of Vavilov in 1940, the house arrest of Kapitsa in 1946, the dismissal of Dubinin in 1959, the ending of the research career of Sakharov and the unemployment of hundreds of Jewish refuseniks prove that political controls on the employment of even the most eminent and seemingly indispensable scientists are an important part of the system. But this is not to say that scientific appointments are entirely outside the control of the scientific hierarchy.

It is not a point I want to make much of. Its personnel powers give the party an enormous degree of control over Soviet science, with much of that control being exercised by the Central Committee apparatus. But it is not total control, and probably not even sufficient to deny the possibility of a scientific community with some sense of its own particular values.

Control functions – political control

Science policy and ideological politics are not easily separable, as should already be clear from the sections on control of policy input and personnel. In this section I want simply to mention some of the other mechanisms the party leadership uses, through the Central Committee apparatus, to ensure that Soviet science, and more specifically, Soviet scientists do nothing that could be in any way seen as threatening the political stability of the regime. We are concerned here with two main issues – control of conferences and overseas travel and control of publishing.

Conferences and travel

As Jewish refuseniks have discovered, control of attendance at conferences and overseas travel are not necessarily synonymous. They were unable to gain entry even to conferences in Moscow, including the International Magnetics Conference in August 1973,[114] while Aleksandr Nekrich, while still on the staff of the Institute of General History, was refused entry to the World Congress of Historians. He states that attendance at conferences is determined, in the case of the Academy, by the relevant department of the Academy, the National Committee of Soviet Scientists, and then confirmed by the Central Committee's Science Department.[115] According to Zemtsov, at the general meeting of the Soviet Sociological Association in July 1972 at which Rutkevich, the purger of *IKSI*, was elected chairman, the delegates were carefully screened, with some being refused admission.[116]

But the most notorious examples of political control of conferences concern attendance at conferences abroad and the travel they entail. Permission to travel abroad (for any reason, including personal) requires, firstly, permission from one's place of work, including a reference from the secretary of the primary party organisation. This requirement covers even non-party members.[117] The application then goes to the *raikom* or *gorkom*, where it is examined and the applicant is interviewed by a special commission. (This applies only to official travel; an application for private travel goes directly to the local visa office after the workplace clearances have been obtained.) One acquaintance who appeared before such a commision related that the members were tough old pensioners who nevertheless asked elementary and rather amusing questions on the fundamentals of Soviet ideology.[118] The application then goes, it seems, to the *obkom* and then one's Academy or ministry, the International Department of GKNT, and then the Central Committee apparatus, both a special 'Departure Commission' and the relevant department. Another 'Departure Commission', of the Ministry of Foreign Affairs, gives a final check, although this takes place when the preparation of travel documents is already well-advanced.[119]

The results of such controls are notorious in the West. Soviet delegations are generally far smaller than Western ones, while delegates are usually older and more likely to be science bureaucrats than researchers. Organisers of international conferences are perpetually frustrated by the failure of listed Soviet delegates to turn up.[120] Soviet ideological authorities are clearly worried about the dangers of sending 'inexperienced' researchers abroad, with the case of Valery Pavlinchuk's trip to Denmark seen as a sad example.[121] Science managers are happy to keep a monopoly of a most valuable Soviet perk, foreign travel. The results are inconvenience for conference organisers, and disaster for Soviet science, this being only one aspect of the general isolation of the Soviet scientific community from international science.[122] We have no way of knowing how important the role of the Central Committee apparatus is in deciding who will or will not attend conferences, but clearly the role does exist.

Publishing
Few need to be told that the Soviet Union has a system of total censorship. It is not so widely known that the censorship function is not left entirely to the formal censorship body, Glavlit. While nothing can be published (and no type-written material sent abroad) without Glavlit approval, most censorship is carried out long before the material gets to the official censors. The system relies to a large extent on self-censorship, but beyond that scientific literature must go through an elaborate system of refereeing, often for political as much as for scientific content.[123] This includes articles for publication in foreign journals. These are not sent abroad by the researcher, but by the institute after checking with other control organs.[124]

The Central Committee apparatus is again involved, with Zhores Medvedev claiming that a party member cannot submit a book to a publisher without prior examination by the Science Department.[125] Both Semen Reznek, the author of a book on Nikolai Vavilov, and Aleksandr Nekrich, writing on the beginning of the Second World War, list the numerous authorities they had to get approval from, including the Science Department, to get their controversial books published.[126]

Again we cannot evaluate the importance of the role of the Central Committee apparatus. In the ideologically non-controversial sciences it is probably not great, although bans on particular authors, including citing their works, have been of more than nuisance value even in the pure natural sciences. The more ideologically sensitive the discipline the more serious these forms of control become and the more likely it is that the Central Committee apparatus plays an important role.

Control functions – policy implementation

Like all party organs the Central Committee apparatus is charged with 'verifying the fulfilment' of the policies and directives of the party leadership. This entails more than simply checking that directives have been obeyed, but includes taking active steps to facilitate general policy implementation.

In the social sciences we have evidence of the apparatus implementing the changes in policy line determined by the highest authorities, for example, encouraging or discouraging sociology. But in the natural or technical sciences we again have very little information. It is said that the relevant Central Committee department checks the draft decisions of ministry collegiums, many of which are concerned with R&D matters.[127] Central Committee representatives attend conferences and meetings at which the latest campaigns are discussed, but they are not usually reported as actively participating, and there is often something inappropriate about the representative, particularly when the Academy is involved. For example, Demichev, an ideologist, attended a meeting devoted to technological progress.[128] Articles in the press, whether the mass circulation newspapers or the specialised journals, by Central Committee workers are rare. But most striking is the lack of reports of the Central Committee apparatus playing an 'expediting' role in research programmes. One might expect the apparatus to have such control over resources that it would be the first point of appeal for researchers unable to get the equipment, facilities and cooperation they need, particularly in the *vnedrenie* of their work. We do in fact have one example. A project on new methods of synthesizing high-isomer carboxylic acids being jointly conducted by the Academy's Institute of Elementary Organic Compounds and institutes of the Ministry of Chemical Industry received close attention from the Central Committee's Department of Chemical Industry and the Committee for Party Control.[129] Since the latter is a party disciplinary body, presumably some heavy pressure was put on reluctant industrial managers. But the rarity of the report, plus the involvement of the Committee of Party Control, suggest the case was exceptional.

The other interesting aspect of this report is that a branch Central Committee department and an Academy institute are involved together.[130] With the Academy being increasingly involved in applied research and large-scale R&D programmes, the Science Department's supervision is clearly becoming less appropriate. It could be becoming easier for the branch departments to involve theselves in Academy research. The Science Department also has to share its responsibilities at the other end of the spectrum, in ideology, with the Propaganda Department.[131] With these pressures on it the Science Department might retreat further to the educational side of its responsibilities. How much these developments are a case of the Academy throwing off the burden of an unpopular Trapeznikov, which might be reversed under a more reasonable Medvedev, is not yet clear.

The apparent lack of expertise within the Central Committee Science Department, and the rarity of reports of its involvement in the natural and technical sciences, are striking. It appears to be almost exclusively an ideological body, perhaps to a large extent cut out of its formal responsibilities for the Academy and GKNT. As for branch research, while the responsibility of the branch departments for a whole branch of the economy, rather than the narrow concerns of a single ministry, might make them more aware of the importance of science and technology than the ministries, the fact is that they are held just as responsible as the ministries for production plan fulfilment, and therefore are subject to the same anti-innovation pressures as industrial managers. When we also remember the small size of the apparatus,[132] scepticism about the involvement of the Central Committee apparatus in science management would seem to be in order. The consequences are two-fold. Firstly, the bulk of the science policy-making process takes place within the state apparatus; secondly, the opportunities are greater for top scientists to supply directly to the highest party leaders the sort of 'staff' advice that would usually come through the Central Committee apparatus. We saw at the beginning of this chapter the importance of the direct links of scientists with political leaders. The findings of the examination of the Central Committee apparatus further accentuates that importance. It lends some support to a Western claim that scientists in general are able to exploit their unique status and 'cachet' to gain access to political leaders without going through established bureaucracies. This access is protected by scientists exaggerating their lack of interest in politics on a broader scale and indeed by presenting an appearance of political naïveté.[133] It seems possible that Keldysh's attacks on *podpisanty* were prompted at least partly by the fear that the overt politicisation of scientists would endanger their special status.[134]

It is not the purpose of this study to examine the science policy-making process within the state apparatus. However some comments are in order. The main actors in the process are the Academy of Sciences, GKNT and Gosplan. They are all formally subordinate to the Council of Ministers, with the chairman of the latter two being deputy chairmen of that body and therefore having seats on its presidium. The Academy of Sciences has no formal representation on either the Council or its presidium. Very roughly, the Academy is responsible for the coordination and conduct of fundamental and social science research, GKNT for applied research, and Gosplan for *vnedrenie*, although these functions vary over time and always overlap considerably.

It is generally considered, in the Soviet Union and the West, that Gosplan's main interest lies in the production plans for which it is primarily responsible. Like the branch ministries for which it writes these plans, it has no particular commitment to technological advance, particularly if it complicates the planning process and threatens even a short-term reduction in gross output. It certainly would not be happy about other bodies being allowed to force changes in its planning because of the needs of R&D. This tends to put it into a natural alliance with the industrial ministries, who of course control and to a greater or lesser extent neglect the branch R&D network or force it into minor 'trouble-shooting' tasks.

Gosplan and the ministries could not fail, therefore, to be disturbed by a number of trends in the last twenty years. The first is the increasing power of the Academy of Sciences and GKNT. Ever since the blow Khrushchev dealt the Academy in 1961 it has been steadily increasing and broadening its functions back into the applied research that it was supposed to have abandoned that year. The most dramatic indicator of its increased role was Brezhnev's statement at the 25th Party Congress that 'the Party values highly the work of the Academy and will raise its role as the centre of theoretical research and the coordinator of all research work in the country.'[135] Previously the Academy was described formally as the coordinator of only the natural and social sciences. The new formulation was definitely seen by the Academy management as an important extension of its powers – by Aleksandrov at the 1976 meeting of the Council for the Coordination of Research in republican Academies, vice-president Fedoseev at the mid-1976 General Meeting, and various other lesser figures.[136] In the following couple of years strong moves were made to increase the role of the Academy in the applied sciences, with, from 1977, a marked increase in the attention given to *vnedrenie* in the Academy's presidium, general meetings and journal. At a General Meeting in

March 1977 a number of changes were made to the Statute of the Academy to reflect the changes in its functions. The technical sciences were included as within the Academy's competence in articles 2 and 4, while article 2 for the first time recognised the supradepartmental (*nadvedomstvenny*) character of the Academy.[137] The Statute was changed again in 1981 to give some power to match the Academy's responsibility for the coordination of all fundamental research. The Statute now set out the Academy's right of scientific–methodological management even of branch institutes when they are involved in fundamental research, while the director of such an institute is appointed and dismissed with the agreement of the Academy's presidium.[138]

At the same time as the Academy was expanding its functions, GKNT was also becoming more of a force to be reckoned with. None of its predecessors had been able to make their mark on Soviet science management, but since its establishment in 1965 GKNT has presented a far more solid face to the world. Its two chairmen have both occupied prominent places in the Soviet scientific world; it is very active in organising the various commissions and councils that are so important in Soviet policy-making; it has reasonably substantial funds to dispose of (about 2 to 3 per cent of the science budget) and apparently some power to direct ministries in the allocation of their funds;[139] it has statutory rights to coordinate all interbranch research; it has the right, which it apparently has used to some extent, to close down poorly performing or unnecessary institutes;[140] and it is responsible, usually together with other institutions, for drawing up the 'methodological' documents that govern so many of the procedures in Soviet research institutes.

Despite the claims of some Western scholars,[141] there is evidence that relations between the Academy and GKNT are quite good. GKNT's two chairmen, Kirillin and Marchuk, have both had very close links with the Academy (unlike the chairmen of GKNT's predecessor organisations); one can find a good number of reports of cooperation and mutual praise; while both organizations reserve their most venomous criticism for the branch ministries.[142] It seems possible to argue that the Academy and GKNT operate as allies in the furtherance in policy-making circles of the interests of R&D, against the more production-oriented interests of Gosplan and the industrial ministries.

What must disturb the production alliance most of all are the indications that the party leadership has a considerable degree of sympathy for the Academy–GKNT axis, at a time when it expresses strong disillusionment with the work of the industrial ministries.[143] The simple fact of the extension of power of these two institutions is evidence enough of this. Particularly with regard to the Academy, the party has shown further signs of favour. This has been most evident at republican and regional level, with republican first secretaries regularly speaking at the General Meetings of their local Academies, and regional party leaders attending presidium meetings at which the work of their region's Academy affiliate is being discussed.[144] The Academy press generally speaks far more of the role of the party in assisting it in its work than ever before. The party shows every sign of relying on the Academy for the implementation of its plans for the technological modernisation of Soviet production. The recent opening in the Academy of a Department of Computing, which was explicitly described as a response to the poor work of the responsible ministries, illustrates this well,[145] as does the strong emphasis in recent years on programme planning, in which the Academy and GKNT both play a major role.[146]

The science lobby in the Soviet Union would thus appear to be in a strong position. Its major institutions have extensive powers over the entire research network and what should be a major initiating role in a totally integrated planning system. They also appear to have the support of a party leadership which is strongly disillusioned with the technologically conservative habits of the branch network.

There are, however, weaknesses in the party–Academy-GKNT coalition. As far as the Academy of Sciences is concerned there is the controversial question of to what degree and how it should involve itself in non-fundamental research. Although there is no evidence of an anti-applied research faction within the Academy as strong as that which was active in the 1950s, there is clear concern among academicians about the degree to which the Academy's institutes have been forced into applied R&D in the last decade or so, much of it on a short-term contract basis.[147] While the Academy leadership seems to be so firmly in favour of an extended role for the Academy that a 1950s type revolt is unlikely, some caution is perhaps still required. Further, and perhaps more importantly, it must be asked whether an institution that receives only about 8 per cent of the science budget and has about 10 per cent of research personnel can solve all the problems of Soviet R&D.[148]

The same question must be asked about GKNT. While it has well-qualified personnel and prestigious leaders it is still essentially a staff organization. It has virtually no reasearch capacity of its own and few operational powers. Continuing proposals that a 'fourth link' of research institutes be set up under GKNT or that GKNT's powers over branch institutes be further extended, or alternatively that R&D funding and resource allocation be centralized under GKNT control, indicate that many feel that as things stand GKNT will never be in a strong position to play a major role in Soviet science management.[149]

This is not just a theoretical question of the right balance between staff and operational roles,[150] but also a matter of bureaucratic clout and politics. The ability of Gosplan to protect its interests even in formal arrangements can be seen in its success in maintaining control of the section of long-term planning that is concerned with *vnedrenie*.[151] One imagines that in informal terms the vast operational powers of Gosplan and the ministries make them truly formidable opponents for GKNT.

The final weakness of the alliance is the hesitancy of the party leadership. The post-Khrushchev leaderships' horror of his impetuous style of leadership, the authority rather than power-based nature of modern Soviet political leadership,[152] and the leaderships' distaste for upsetting established bureaucratic procedures,[153] have led to a timidity in policy-making, particularly in industrial management, that borders on half-heartedness. Despite the bitter public criticism by party leaders of the industrial ministries, the results in terms of new policies and approaches have been so intricate and incremental that non-implementation is virtually guaranteed. The picture in research management is much the same. The enemy has been identified, but the party leadership fears to do more than skirmish. Decrees on science management call on the Academy to play a greater role without significantly increasing its powers or its funding,[154] and repeat the same calls for minor changes.[155] Those new systems and approaches that do gain some degree of formal implementation are extraordinarily complex and bureaucratic, particularly material stimulation and performance and economic evaluation schemes, while more dramatic proposals from science management specialists and practitioners are ignored. A degree of scepticism with regard to Khrushchev's approach is understandable. But given the party leadership's clear concerns and apparent confidence that the Academy, in particular, and GKNT can help save the situation, one might have expected more resolute action. The question becomes whether the lack of action is a deliberate, rational policy choice or the result of a fundamental weakness of the modern Soviet political system.

There appears to be no reason to believe that science policy-making is any different from any other aspect of policy-making in the Soviet Union, in that the top party bodies, specifically the Politburo, are ultimately responsible for all major decisions. Clearly scientists have considerable opportunity to provide input into the policy-making process, both through informal contacts with top political leaders and through more institutionalized channels. These institutionalized channels apparently include the Central Committee apparatus to a lesser degree than might have been expected. The apparatus appears not to have the capacity to play a major role in either science policy-making or management.

Despite this, the Politburo has ample and reliable input from scientists and from the various competing institutions in the state apparatus. Among these institutions the Academy of Sciences and GKNT seem to have considerable influence, with the party leadership publicly adopting their point of view rather than that of what might be seen as the opposition, Gosplan and the ministries. However, the end results, in terms of policies both as issued and as implemented, are weaker than one might have expected from the rhetoric. The reasons for the weakness are fundamental to our understanding of the Soviet political system. The dominance of the party leadership in policy-making has led many to see the Soviet political system in 'vanguard party' terms. But the timidity of the results in present circumstances might lead one to question that finding.

3 The Regional Party Apparatus and Science Management[1]

The USSR is a sufficiently centralised state that one does not expect a great deal of independent policy-making outside the top-level political bodies in Moscow, even at times, as at present, of considerable emphasis on territorial or regional planning and management of the economy.[2]

This is not to deny local organisations considerable powers in interpreting and modifying for local conditions policies which have been centrally set. While the Soviet Union is, on the one hand, notorious for its inflexibility in making allowances for local conditions, particularly in agriculture, on the other hand a number of Western observers have fruitfully applied to it the *de facto* decentralising effects of 'conflicting standards analysis'.[3] This has led to the description of the functions of regional party bodies in terms of a 'prefectural' or 'brokering' role, best seen in the work of Jerry Hough.[4]

But the main task of the regional party organs is the faithful implementation of the policies set by the centre, both furthering general priorities, such as the stress on scientific and technical development since 1955, and ensuring the fulfilment of specific plans and directives.

These functions of the middle-level party apparatus will be examined in this chapter under two main headings – their control of and assistance to individual institutes in their day-to-day research, and their promotion of innovatory measures on a broader scale. The difficult task of assessing the effectiveness of their involvement will then be attempted, followed by an examination of the machinery available to party organisations for science management, especially the background and training of the party workers involved.

The involvement of the regional party apparatus in the control of scientists and scientific institutions has never been negligible. From the earliest days of Soviet rule, regional party organs have been held responsible for those most important of party tasks, ideological control and personnel management, in scientific institutions. At times of campaigns of particular political importance their role seems to have increased. For example, local party organs were directly involved in the establishment of a party presence in the Academy of Sciences in the late 1920s,[5] and the recruitment of suitably Bolshevik proletarians to science during the 1920s, particularly the *vydvyzhenie* movement (a worker–peasant affirmative action programme taken to extremes) of the First Five Year Plan.[6] According to Soviet sources the links between the regional party apparatus and scientists became particularly close during World War II,[7] while the former played a major role in the campaign to increase party membership among scientists after the war.[8]

Nevertheless, particularly in the supervision of research work which was not ideologically sensitive and in general science management, the party's role was not great.[9] It is interesting that there is no evidence of party involvement below the very top level in such major research programmes as aircraft construction and the development of the atom bomb. Presumably the fact that a lot of such research was done in prison laboratories is of some relevance.

The ascendancy to power of Khrushchev in the mid-1950s brought a potentially major change, given that the new leader had a strong commitment to science and a policy of an increased role in the economy for the party apparatus. Khrushchev brought about an upsurge in the involvement of the party apparatus in the day-to-day operations of research and design institutes. One of the first major articles marking this upsurge was published by a secretary of the Sverdlovsk *obkom* in the final issue of *Kommunist* for 1954. It was devoted to the newfound concern of the *obkom* with the research institutes of the Urals branch of the Academy of Sciences.[10] In July the following year a Central Committee plenum devoted considerable attention to shortcomings in industrial research, with the party apparatus being directed to involve itself more closely in improving research work.[11] The plenum was followed by a number of articles on the role of the party, with the stress on working with primary party organisations.[12]

As we will see in the next chapter, the primary party organisations in applied research institutes were granted new powers in 1959, with permission to establish control commissions, and in 1961 to use the right of control. A feature of the new powers and responsibilities of the party apparatus was involvement in the day-to-day running of research. The primary aim was the timely introduction of research results in production, rather than solely the traditional concerns with ideological conformity and routine plan fulfilment.

The effort was not sustained in the final years of Khrushchev's rule. The first half of the 1960s was a period of the closest involvement of the party apparatus in the minutiae of economic management and above all in ensuring the achievement of the most important Soviet goal – the fulfilment of production plans. The November 1962 Central Committee plenum strengthened these pressures by dividing the party apparatus into industrial and agricultural wings. In such circumstances anything not immediately connected with production management, including scientific research, tended to be neglected. One suspects that in this period the regional party apparatus saw science as no more than a distraction from the more immediate and more important problems of production.

The fall of Khrushchev saw a significant change in the situation. The original impetus for change, however, was not connected with research itself, but rather with the need to re-establish control over the political situation in research institutes, something which had got somewhat out of control in the final years of Khrushchev's rule.

As we saw in a previous chapter, the first year or so of the new leadership was a period of relative instability in the ideological area, and, if anything, of greater liberalisation than had been evident at the end of Khrushchev's reign. But by the time of the Czechoslovak crisis in mid-1968, the party leadership was ready to take measures against dissident scientists. In fact the measures required to bring widespread dissidence among scientists to an end were relatively mild – the expulsion from the party and dismissal from their jobs of a representative sample of *podpisanty* (signers of protest letters and petitions) seemed to be enough. Local party authorities played a major role in these measures; the indications are that they were held particularly responsible by the central leadership for the state of ideological discipline in Soviet science from the late 1960s. Zhores Medvedev describes the role of the head of the Ideological Department of the Obninsk *gorkom* (one is unsure whether he means the Agitprop Department or the Ideological Commission) in sacking the head of the local newspaper and generally leading the crackdown on dissidents in that city. The purge in Obninsk apparently included party officials, with one *gorkom* secretary being sacked for 'spinelessness' and replaced by a tough *apparatchik* from a rural *raikom*.[13] Rudol'f Ianovsky, the first secretary of the Sovetsky *raikom* in Novosibirsk, in 1970 related with some sense of satisfaction the success the *raikom* had had in reestablishing order in Akdemgorodok,[14] while *samizdat* publications describe the role of *raikomy* and *gorkomy* in taking disciplinary measures against *podpisanty*, often in the face of the reluctance of primary party organisations.[15]

The relative political peace in research institutes since the early 1970s has reduced the role of the regional party authorities in this area.[16] However the end of dissidence as a mass phenomenon coincided with another development which ensured that the party did not drop out of science. This was the time when Brezhnev began to assert his superiority over the rest of the members of the Politburo. This had two consequences of relevance to our discussion. Firstly, the dominance of Brezhnev, the party *apparatchik*, brought about an ever-increasing reliance on the party apparatus in the management of the economy. While Brezhnev's typically cautious approach meant that the apparatus was kept under control, with periodic minor campaigns against *podmena* (the excessive involvement of the party in day-to-day management of the economy and production), on the whole the last ten years or so have been marked by the swing of the pendulum very much to heavy party involvement in day-to-day economic management. In earlier times this could well have led to a neglect of science, as short-term production plans became the sole concern of party secretaries. However, the second consequence of Brezhnev's rise to dominance was the acceptance of the scientific-technical revolution as one of the guiding principles of Soviet economic development. This meant that the party apparatus was expected to regard science and technology as at least approaching the importance of production plans.[17]

There is evidence that the expectation has to some extent been realized. Certainly some of the more important research areas show a high degree of party interest in terms of plenums and secretariat and buro decisions. In the second half of the 1970s the Novosibirsk *obkom* examined the work of an institute or a major R&D question three or four times a year.[18] Between 1972 and 1974 the Lenin *raikom* in Moscow held eight plenums on scientific and technical progress, six meetings of the party *aktiv* and 34 meetings of the buro.[19] There are two aspects of an increased role for regional party organs in science management. Firstly, there is their involvement in the day-to-day direction of research, working in the institutes just as they always have in the factories and farms, egging institute management on to fulfil research plans and helping through their extensive range of contacts to solve all the problems of communication and supply that arise in any part of the Soviet system. Secondly, there is evidence of party apparatus involvement in the larger issues of the reform of the whole R&D system, with the party's role seemingly to provide a progressive push against the obstructionism of entrenched interests. The next two sections present evidence of regional party involvement in these two areas. No claims are made at this stage of the nationwide importance or applicability of the phenomena described. An attempt will be made to put them into perspective in a later section.

INVOLVEMENT IN DAY-TO-DAY RESEARCH

Regional party organs have always been willing to turn to local research organisations to help solve operational production problems. For example, in the mid-1960s the Rostov *obkom* looked to local scientist to solve agricultural problems.[20] But with the coming of the scientific–technical revolution, as Shevardnadze, then Georgian party head and candidate Politburo member, made clear in 1982, such a narrow approach is no longer adequate.[21] Party leaders now have to devote particular attention to R&D and science management.

Having been made more responsible for technological development in their localities, regional party committees have exhibited a new desire to help research institutes carry out their research and particularly implement the results in production. If the enterprise to receive the new technology is in the locality the party committee will keep an especially close eye on the progress of the R&D work being done.[22]

A situation of scarce resources means that the local party organ will immediately be faced with a problem of setting priorities. Which sort of research and which institute are of most importance for regional interests, or are understood by the local party secretary to enjoy high priority in the eyes of higher authorities? It seems to be recognised as a legitimate and indeed necessary task for regional party organs to set research priorities on a regional basis. In 1982 Shevardnadze called on local party organs to determine the most important local scientific problems, and even to use formally organised competitions (*konkursy*) to solve them,[23] while it is claimed that the giant '*Sibir*" programme drawn up by the Siberian department of the Academy of

Sciences for the development of Siberia was checked by the Novosibirsk *obkom* and Sovetsky *raikom* for relevance to the needs of the area before being approved by those bodies.[24] Within the framework of the set priorities the party organ is expected to coordinate the work of all research organisations within its territory, and obtain the cooperation of organisations outside its territory.[25]

Once priorities have been decided, the most usual type of assistance required from party organs is in getting equipment and facilities, and then finding a ministry or production enterprise willing to either develop or produce a new product or process. Regional party authorities are used to struggling with the notorious inefficiencies and tardiness of the Soviet construction industry, and are able to apply their experience in this area to high priority research institutes. Thus the Sverdlovsk *obkom* was closely involved in the construction of the Urals Science Centre in Sverdlovsk, going so far as to in one instance 'order (*obiazat'*) the Chief Central Urals Construction Administration to continue a road, sewage system and other engineering works to the Gorelevsky Boundary [the site of the science centre].'[26] Similarly the Irkutsk *obkom* considered it its duty to end the construction delays that were hindering R&D work and implementation in its region.[27] Sliusarenko describes how the L'vov *obkom* 'compiled precise timetables for the construction and handover of scientific complexes', and how on the *obkom*'s initiative Student Construction Brigades were organised to help the local construction trust put up buildings for *vuzy* and research institutes.[28] One third of the eighteen decrees and decisions of the Novosibirsk *obkom* on R&D matters listed by R. G. Ianovsky are on construction questions.[29]

In recent years there has been heavy stress on building up experimental and prototype facilities for research institutes, as a solution to the traditional Soviet problem of poor introduction of the results of research into production. There is evidence to suggest that at times regional party organs are expected or forced to involve themselves in finding spare industrial capacity that can be transferred to institutes. Thus in 1966 the Rostov *obkom* helped in finding experimental facilities for the All-Russian Research Institute for the Mechanisation and Electrification of Agriculture in Zernograd and the Rostov Institute for Machine Building Technology.[30] The problem is a particularly difficult one in the metropolitan areas, such as Moscow, Leningrad and Kiev, where there are strict limits on new construction and importation of personnel.[31] These limits lead to disputes over whether inner-city industrial facilities should be handed over to institutes. In Leningrad in 1966 there was clearly some policy dispute over this question. The head of the Industry Department of the Leningrad *obkom* was opposed to industrial facilities being granted to institutes; indeed he considered that design bureaus that had in the past been hived off from factories should now be given back to them. However when the *obkom* eventually made its recommendations on the question, provisions were made for the transfer of industrial facilities to institutes.[32]

Once facilities have been provided, research institutes are always likely to face problems with equipment. Soviet researchers themselves build up considerable expertise in ferreting out and bargaining for scarce equipment. However local party bodies claim to help here as well. The Cheremushki *raikom* of Moscow establishes contact between enterprises and institutes of different branches within its *raion* in order to improve the supply of equipment.[33] Regional party bodies also claim to be involved in attempts at setting up institutions designed to improve the scientific technical supply problem. Thus in 1966 the Leningrad *obkom* examined criticisms of the supply of equipment to research institutes, as a result of which a special organisation, 'Leningrad Scientific Supplies', was set up to open shops for over-the-counter sales of equipment. In 1973 the *obkom* recommended to the Chief Supply Administration of the Leningrad economic region that it organise a system for the hiring of equipment.[34] As usual in such cases there is no reason to believe that the initiative was necessarily the *obkom*'s. However the approval of the local party authorities would be considered the minimum necessary condition for any such schemes to be actually implemented.

Most party secretaries claim that it is not necessary for the *raikom* or *obkom* to involve itself in the details of the research process itself. As the first secretary of Moscow's Chermushki *raikom* says:

> There is no need for the *raikom* to enter into the specifics of all research projects, into the essence of the different directions of research. After all, the primary party organisations, party committees and buros of each scientific institution are obliged day-by-day to concern themselves with these questions within the limits of the powers granted to them.[35]

In the next chapter I show the great increase in the involvement of institute primary party organisations in research matters since the late 1960s, particularly such questions as personnel appointments and movements, weeding out unwanted projects, labour management techniques and productivity schemes. Nevertheless, examples can be found of *raikom*, *gorkom* or even *obkom* involvement in essentially intra-institute matters, although very often these are concerned with 'the spreading of leading experience'. Thus the industry–transport department of the Smol'ny *raikom* in Leningrad worked closely with the State Union Design Institute (GSPI) to develop and recommend to others a system of alternative design (*variantnoe proektirovanie*),[36] while the Cheremushki *raikom* in Moscow ensured that a common system of performance evaluation was used in institutes throughout the *raion*.[37] The regional party bodies are also forced to step in in cases where the primary party organisation is unable or unwilling to take action against poor or illegal management procedures.[38] One area where there is no doubting the direct involvement of regional party bodies in the internal affairs of research institutes is in personnel control. According to emigre sources the appointment of department and laboratory heads and even promotions to senior research worker are subject to *raikom* or *gorkom* confirmation.[39] The first secretary of the Smol'ny *raikom* of Leningrad has stated that the leading executives of all research and design organisations, the secretaries of party committees and buros and a range of scientific and technical specialists are on the *raikom*'s *nomenklatura*.[40] In 1980 the secretary of Moscow's Cheremushki *raikom* claimed that 33 per cent of the decisions of the *raikom* buro on science matters were connected with personnel.[41] However at this level the same questions about the actual procedures of *nomenklatura* control arise as arose with regard to the Central Committee *nomenklatura*. To what extent does the party organ provide the initiative in personnel appointments, or does it rely primarily on recommendations from leading scientists? Who is likely to win in case of conflict?

Except for these special cases local party involvement in the actual scientific process is rare, and it seems probable that local party organs rely primarily on primary party organisations for direct action in these areas of concern. There is not the evidence of the constant presence of *raikom* representatives actually within the institutes that one can find in enterprises and farms.

Greater regional party involvement is likely to become necessary again when the time comes to find industrial sponsors for a particular piece of completed research. The shortage of experimental facilities in Soviet research institutes means that often development work has to be done in industrial enterprises. Then, even when a proven prototype is available, a ministry or enterprise has to be persuaded to take on production of the new product. Again there is evidence of efforts by regional party organs to use their influence and contacts to help research institutes with this problem. For example, the All-Union Research Institute for the Technology of Electrical Machine Construction has an experimental factory, but it can produce a new product only one at a time. Kharkov *gorkom* therefore took it upon itself to persuade the Ministry of Electrotechnical Industry to mass produce a particular new lathe.[42]

Concern with the links between science and production is the aspect of R&D which appears to receive the most attention from regional party bodies. They use direct contacts with the relevant research and production organisations or encourage the various people involved to get together at conferences, congresses and councils organised under the auspices of the party committee. As a reasonably formal device, considerable use is made of the so-called *dogovory sotrudnichestva* (contracts of cooperation). These have become very popular in recent years, apparently for two interconnected reasons. Because they are not legally binding documents but rather informal agreements between partners they are more open to influence from local party authorities. Also they are more flexible than the bureaucratically stifling 'economic contracts' (*khoziaistvennye dogovory*). They are apparently sometimes little more than agreements to exchange information and laboratory techniques and to share expensive apparatus. However they can also cover the sharing of research projects and, more particularly, agreements for development and series production of new ideas and products. Local party organs clearly put considerable pressure on research institutes and especially industrial enterprises to sign such contracts.[43]

Another related aspect of the party's concern with *vnedrenie* is its sponsoring of independent inventors. Often these inventors in fact work in a research institute, but have trouble getting their ideas accepted by their peers. For example, a professor from Mogilev had an idea to use wooden components in the rollers on conveyor belts, but was unable to get the support of the relevant branch institute to have his idea tried out. While in Mogilev he had had the support of the *gorkom* and some progress was made. However he had recently moved to Petrozavodsk. Despite all the relevant information being sent to the local *obkom*, the party organisation showed no interest in his idea and there seemed to be little chance of its implementation.[44] A 1981 article in *Kommunist* deals with a number of similar cases, including the Gorky *obkom* supporting a couple doing research on providing a natural atmosphere within spacecrafts despite the opposition of scientific colleagues. The author of the article is confident that 'in general the level of party cadres in Soviet science is such that they are able with adequate competence to involve themselves – either independently or with the use of consultants – in complex scientific and technical problems, and then with initiative and creativity, considering all aspects of intellectual production, arrive at authoritative conclusions. This broadens the possibilities of party involvement in scientific relations, particularly when ethical considerations arise in the conflict of interests.'[45] Despite this confidence one wonders how often inexpert and gullible party workers are taken in by scientific charlatans or cranks. Lysenko was such a one who succeeded on a grand scale. As mentioned in a previous chapter it is considered to be one of the functions and virtues of the party to act as a 'vanguard' in all areas, willing to listen to and adopt radical ideas, overcoming in the process the conservatism and inertia of the established bureaucracies. It is a role that can lead to laughable and sometimes tragic results. Nevertheless it does have possible benefits, in leaving some room for an entrepreneurial spirit.[46]

I do not wish to suggest that in any of these areas – construction, equipment, seeking industrial sponsors – the regional party's role is unique or decisive. The Soviet R&D system has its own well-developed procedures, both formal and informal, for all this. The experienced scientist–administrators who occupy most senior positions in Soviet science have a considerable talent for this type of work. Nevertheless, as has been copiously documented in numerous Western and Soviet publications, the system does not work well. The regional party apparatus is well placed in terms of the influence and contacts it has to help overcome some of the problems. The evidence suggests that since the late 1960s the party apparatus may have been more willing than ever before to use its position to assist research institutes in their day-to-day work.

INVOLVEMENT IN REFORM OF THE R&D SYSTEM

In the same period there is evidence of the involvement of the regional party apparatus in some of the reforms of the Soviet R & D system that have been so haltingly introduced. This is of particular interest because it represents a good example of the 'vanguard' role the party apparatus is still expected to play in Soviet society. Of particular interest to us is the regionalisation of science management in recent years. This is in the context of an increasing emphasis on regionally-based planning and economic development in general. To a large extent a consequence of the development of the remote regions of Siberia and the Far East, it is also an indication of the intention to give an increased role to the party apparatus in economic management. Being the only regionally organised institutional structure of any significance in the Soviet Union, the party apparatus stands to gain most from a regional emphasis in economic development. If this regionalisation were found to represent a genuine change in the Soviet approach to economic management it could mark a major shift in the balance of power between the various sectors of Soviet society, particularly the party apparatus and the branch ministries.[47]

Science has also been subjected to the regional approach. One of the earliest and best-documented examples is in the L'vov *oblast'* of the West Ukraine. The system of scientific organisation, based on the Western Science Centre which has been in operation there since 1976, was examined in detail by the presidium of the All-Union Academy in late 1980.[48] The discussion, after recording that the work of the Ukrainian Academy had been praised by Brezhnev at the October 1976 and November 1979 Central Committee plenums and at the meeting of the presidents of the Socialist Academies in 1977, concentrated on the inter-branch scientific–production associations and complexes set up under the Bureau of the Western Science Centre of the Ukrainian Academy of Sciences. These are groups of research institutes, design and experimental production facilities, and industrial enterprises linked together on a contract basis. They are headed by a leading scientist, recommended by and usually a member of the Bureau of the Western Science Centre. Planning is done on a regional level, while financing is organised by the Ukrainian Academy with the participation of GKNT. The discussion attracted considerable interest from leading members of the Academy, and following a concluding speech by Aleksandrov, the president of the Academy, describing the experiment as a valuable one, a motion was passed that it be adopted by other scientific centres.

Of particular interest to us at this stage is the very large role given the regional party apparatus in the system. It was claimed by Boris Paton, president of the Ukrainian Academy of Sciences, that the whole experiment was begun on the initiative and operates with the constant participation of the *obkom*. The deputy chairman of the Western Science Centre stated that the plans of the associations' work are drawn up under the general leadership of the *obkom* and that the management personnel of the associations are confirmed in their posts by the *obkom*. He declared that the head of the relevant *obkom* department is deputy head of each association. The first secretary of the L'vov *obkom* addressed the meeting of the presidium, while Aleksandrov in his concluding speech acknowledged that the direct involvement of the party was the driving force behind the experiment.

Soon after, in January 1981, a seminar was organized in L'vov by the All-Union and Ukrainian Academies, the Western Science Centre and the L'vov *obkom*, to share the experience of the L'vov system with other science managers and party officials.[49] In opening the seminar Kotel'nikov, vice-president of the All-Union Academy, declared the experience to be gained from the L'vov experience to be valuable, and to have revealed one general principle, that 'where the regional party apparatus is involved things go well'. The first secretary of the L'vov *obkom* again addressed the seminar and revealed that in 1979 the Ukrainian Central Committee had issued a decree praising the L'vov system.

Similar systems to the L'vov one have since been adopted in other regions. One of the best documented is the Inter-branch Coordinating Council of the Soviet Northwest, recently reorganised as the Leningrad Science Centre.[51] Although based in Leningrad the Centre covers R&D activities in a large area, including Leningrad city and *oblast'*, Arkhangel, Vologda, Murmansk, Novgorod, Pskov *oblasti* and the Karelian and Komi ASSR. The old Council had 36 members (scientists, party officials and industrial managers) and directed the work of 14 specialised scientific councils. The councils draw up R&D programmes covering all work in the organisations involved, from the basic theoretical research to start-up in mass production. The programmes are fitted as far as possible as integrated units into the All-Union R&D programmes drawn up by the Academy of Sciences, GKNT and Gosplan. The role of the regional party apparatus appears to be great. Eight of the 36 members of the old Council were party officials, while programmes are worked out in cooperation with advisory committees operating within the *obkomy*. With the Council for Energy and Electrical Power Machine Construction as an example, it was stated: 'This council uses an entirely new form of work, in which the power and authority of party leadership and the authority of leading energy researchers are organically combined.'[52]

The involvement of the party apparatus is not just for show. Scientists such as Paton make it very clear why they welcome the party's involvement. In discussing regional R&D programmes and the six scientific centres in the Ukraine, of which the L'vov centre is one,

Paton announced in 1981 that the presidium of the Ukrainian Academy had concluded contracts with all *obkomy* in the republic and the Kiev *gorkom* and that 'these contracts cover scientific research work within the boundaries of the region . . . . The point is that in conditions where many industrial enterprises and construction organisations are subordinate to union ministries, only action from the party makes it possible to overcome departmental barriers'.[53] Few details are available on what these contracts cover. Paton says of them unhelpfully:

> These contracts, which have become one of the factors in the acceleration of scientific and technical progress, cover a complex of measures for the intensification of the activity of industrial enterprises situated in the region. They are usually preceded by considerable preparatory organisational work, in the process of which the range of problems which are of interest to producers and scientists is uncovered. The day-to-day management and control by the *obkomy* help the successful solution of the scientific and technical tasks set out in the contracts. Each contract is set out in the form of a document signed by the Academy's management and the first secretary of the relevant *oblast'* committee. Those actually responsible for fulfilling them are, on the one side, the Academy institutes, and on the other, individual enterprises and production associations of the *oblast'*.[54]

But there is no doubt as to their main purpose – to declare an alliance between the party and scientists, primarily those based in the Academy, against the notorious villain of all attempts at economic and administrative reform in the Soviet Union, the branch ministries.

In a similar vein, when the All-Union Academy presidium was considering the work of the L'vov system, the deputy chairman of the Western Science Centre was asked what happened when an attempt was made to introduce new technology which was not profitable for production enterprises. He answered laconically that the party organs help.[55] The chairman of the Interbranch Coordinating Council of the Northwest stated in connection with the Leningrad scheme:

> The participation in development and realisation of relevant goal-oriented programmes of the party's *raion* committee and the industrial departments of the city and *oblast'* committees of the CPSU has meant that the managers of industrial enterprises can no longer allow themselves to play the role of bystanders in the preparation of research plans.[56]

The dissatisfaction of the party apparatus with the autarchic 'departmental' (*vedomstvennye*) tendencies of the branch ministries has become so strong that one can even read calls by regional party leaders, including the first secretary of the L'vov *obkom*, for the reestablishment of *sovnarkhoz*-type regional bodies.[57] In the field of science this dissatisfaction is shared by the Academy of Sciences, and quite possibly by GKNT.[58] Thus, as was described in the last chapter, the party leadership has declared the Academy to be the leading force for technological progress in the Soviet Union. By encouraging the regionalisation of science management the leadership has made the regional party apparatus a key support for the Academy.

APPLICABILITY AND EFFECTIVENESS

We have now looked at the two levels of the involvement of the regional party apparatus involvement in R&D – as part of a movement for progressive change on a broad scale and as a source of assistance and influence for individual institutes. However the developments described here have taken place on a rather limited scale. How generally applicable to Soviet science as a whole are they, and how effective are they likely to be if more widely applied?

To begin with the regionalisation of science management, it would be wrong to exaggerate the importance of the regionally-based scientific programmes on which I have concentrated here. Nevertheless, they have shown major growth and development, and have received considerable high-level attention. When the Western Science Centre began its L'vov based programmes in 1976, 442 people worked on them, with a budget of 600000 roubles at their disposal. By 1983 the staff had risen to over 9000 scientific workers in institutes and *vuzy*.[59] A large number of other science centres have now been set up, with the Ukraine having six. There have been calls for new ones outside the Ukraine,[60] with the All-Union Academy having four.[61] The existence of such centres provides the framework for regional research programmes controlled by the local party organisation.

This is particularly so if the science centres are seen as part of an Academy–party alliance. While there have been some slight signs of opposition to L'vov-type programmes,[62] they have received high-level support, including praise from Brezhnev himself. Brezhnev's death has not meant the end of such concerns. At the first post-Brezhnev plenum of the Ukrainian Central Committee, Shcherbitsky made a point of reiterating the interest of the central party authorities in the R&D programmes of the Ukraine. He said: 'Realising the importance of these questions, the Politburo of the Central Committee considers it essential that progress in the fulfilment of the indicated programmes be examined at a plenum of the Central Committee of the Communist Party of the Ukraine.'[63] Both Paton and the first secretary of the L'vov *obkom* spoke at the resulting plenum, presumably on the L'vov R&D system. The establishment of the Leningrad Science Centre in more recent times is an indication of a continued commitment to such an approach. It is also evident in the recent Central Committee decrees on the Urals Science Centre and science management in Leningrad, as well as an enormous academic literature.

Of course the spread of such methods throughout the Soviet Union is hardly a guarantee of their effectiveness. It is beyond the scope of this chapter to determine the pros and cons of a regional approach to R&D. One could well ask whether the linking of fundamental research institutes to production, which the L'vov system is designed to promote, might not become excessive, and whether indeed the approach might not lead to the disruptive regional autarchism of the *sovnarkhoz* period.

Is there any evidence that those areas in which the regional party apparatus is closely involved are more successful than other areas of the Soviet Union? One has to beware of the opportunities for self-promotion available to those with good party contacts, but the impression to be gained from the Soviet press is that the Ukraine, with apparently the strongest links between scientists and party officials, is one of the leading areas of Soviet R&D. The success of an institute like Paton's Institute of Electrical Welding, the director of which has close and well-publicised party links, seems to be undeniable. The increase in scale of the Western Science Centre's system in L'vov and the enthusiasm of the party and Academy of Sciences leadership for such systems might suggest that some degree of success has been had. Leningrad, the other centre of the approach, is certainly accepted by many, both in the USSR and the West, as one of the Soviet Union's R&D success stories.[64] Those involved in running the systems are generous in the praise they give the local *obkomy* for making that success possible.

However a couple of points which suggest some possible weaknesses of the regional programmes need to be brought out. Firstly, they are based on already existing research and production organisations which usually work together on the basis of 'contracts of cooperation' or on a voluntary basis (*'na obshchestvennykh nachalakh'*). These contracts are not legally binding, and work under them has to be fitted into the regular research and production plans of the organisations involved. If, as it appears, they are funded from central Academy and GKNT contingency funds, the organisations involved might be prepared to fit them into their plans. Nevertheless, as is made perfectly clear in Soviet sources, science centres have no powers over the ministries on which they rely to such a large extent for *vnedrenie* and even much of the applied research they do. It is for this reason that they rely so enormously on local party organs.[65] Some participants are confident that the power of the regional party apparatus is sufficient. Academician Zuev, who runs the Tomsk filial of the Siberian Academy, writes: 'For the major complex goal-oriented programmes which are worked out by the sections, relevant decisions are taken by the buro or secretariat of the party *obkom*, and consequently for all participants in the cooperative endeavour, regardless of their branch subordination, fulfilment of the programme becomes essential.'[66]

Nevertheless there are signs that the party apparatus is not always easily able to get its way. Ministries and Gosplan are regularly criticised for their neglect of R&D programmes at all levels,[67] and the Academy is becoming more and more involved in all stages of the R&D process, apparently because it is becoming increasingly difficult to get many of the branch ministries to devote themselves to it wholeheartedly.[68] The experience of the Urals Science Centre, which has suffered long construction delays despite even a Central Committee decree, is instructive.[69] The calls of some regional party leaders for new regional economic management bodies with real legal powers suggest the frustration some of them feel under present circumstances.[70]

When one looks at the record of regional party organs in the more mundane aspects of their work with research institutes, one becomes even more sceptical of their ability to play a major role in R&D. It is noticeable that the number of complaints about the indifference of regional party organs to the work of research institutes and particularly their primary party organisations has fallen significantly in recent years, and there are signs of some regional party leaders genuinely realising the benefits R&D can bring to their regions. The party leadership of Ivanovo *obkom* appears to be particularly keen to get Academy involvement in the technological rejuvenation of the textile industry. The *obkom* first secretary invited Aleksandrov to Ivanovo in 1978 and soon after called for the establishment of a science centre in Ivanovo headed by an Academy institute. Such an institute was set up in 1981.[71] Other local party leaders have made similar requests of the Academy.[72] They have a ready forum to do so since they have begun attending Academy presidium meetings which are considering research in their localities.

Nevertheless one still suspects that for most local party leaders technological development does not come high on their list of priorities. Despite the leadership's emphasis on scientific progress, it is probable that the continuing domination of physical output and other production-oriented indicators in the measurement of Soviet economic performance still has an important influence on the priorities of local party leaders. There would not seem to be much to gain in terms of a party career from a commitment to science. It is true that Dobrik, the first secretary of the L'vov *obkom*, seems to have been rewarded for his support of innovation in research management with candidate membership of the Ukrainian Central Committee Buro and full membership of the All-Union Central Committee. There are also the occasional practising scientists who move over to party work with some success. Boris Chaplin was a State Prize-winning mining engineer and *vuz* lecturer who went on to be first secretary of Moscow's Cheremushki *raikom* and Soviet Ambassador to Vietnam. K. M. Bogomiagkov is a Lenin Prize-winning geologist who plays a very public role as first secretary of the Tiumen' *obkom*. Those whom one might call 'technocrats' have done well in the Leningrad party apparatus, with Romanov's successor as *obkom* first secretary, L. N. Zaikov, having no previous party experience and having been earlier in his career the general director of an unnamed and most probably defence-oriented research enterprise. He has now replaced Romanov as Central Committee Secretary for defence matters.[73] But these cases would not seem enough to overcome the tradition that regional party leaders base their careers on a production background, and the commitment of the local party apparatus to traditional indicators of success. One would therefore expect them to have sympathy with the reluctance of local industrial managers to jeopardise plan fulfilment through involvement in risky innovation.

But more striking than this is the evidence of the lack of success enjoyed by local party leaders in promoting research in their regions even when they apparently approach the task with enthusiasm. The Rostov *obkom* had been unable to obtain experimental facilities for one enterprise despite ten years of negotiations with *sovnarkhozy*, State Committees and ministries;[74] the Kharkov *gorkom* was trying in 1972 to persuade the Ministry of Electrotechnical Industry to mass produce a new product, but without success;[75] despite the efforts of the Irkutsk *obkom*, construction delays and the postponement of factory modernisation prevented fulfilment of the second year of the Five Year Plan for new technology in 1973.[76] The *Pravda* correspondents who wrote of construction delays in the Urals Science Centre in 1974 specifically excluded the Sverdlovsk *obkom* from blame. The *obkom* often considered the problems of the centre and did everything it could to help. Nevertheless serious delays continued. The fact that the *obkom* was not held responsible indicates that it was not expected to be able to solve the problems, despite its very best intentions.[77] Koptiug, the head of the Siberian Academy, acknowledges the important role of the regional party apparatus in encouraging direct links between Academy institutes and production enterprises, but also claims that such cooperation works *only* when the relevant ministry is vitally interested, regardless of the attitudes of the local party leaders.[78] The findings of the recent US survey of emigre R&D personnel also point to the lack of an effective role of regional party organs in the science field. Only a small proportion of respondents, around 15 per cent, saw local party organs as important for their institute's research programme. However a higher percentage – 44 per cent among those working in *vuzy*, 40 per cent in Academy institutes, 28 per cent in branch institutes, although only 17 per cent in defence institutes – had turned for assistance to party bodies outside their institute, which would indicate higher expectations of success. But when asked about which problems they appealed to party organs about, 'others' came far higher than business matters, the resolution of personal conflict, the supply of equipment, solving bureaucratic problems or promotion. Presumably, therefore, most contacts were over personal matters not connected with work.[79]

If regional party organs are unable to solve specific problems for individual institutes and researchers, one wonders how they can cope with forcing through major and wide ranging reforms. One suspects that in science, as in other areas of the economy, big industrial ministries are powerful enough to obstruct even determined regional party organs. When one looks at the resources the party organs have available perhaps this is not surprising.

THE SCIENCE MANAGEMENT MACHINERY OF REGIONAL PARTY ORGANS

It is not an easy task to sort out what parts of the regional party apparatus are responsible for science management. Indeed the administrative arrangements of regional party organs are in general rather incoherent, with a mixture of functional and specialised branch approaches.[80] Thus each regional party committee has, for example, a secretary for ideological affairs, with an Agitation and Propaganda Department under him. This secretary and his department have overall responsibility for ideological questions throughout the region, but will also have general responsibility for so-called 'ideological' institutions, such as newspapers and theatres. Similarly the secretary for industry and the Industry Department under him are broadly responsible for the state of industry in the region, but also have specific responsibility for all aspects of party work in industrial enterprises. An instructor in an Industry Department is made responsible for a group of factories and their party organisations. According to a Soviet handbook, if a non-industrial question comes up, he should not claim that it is outside his area of responsibility. While he is allowed to consult with apparatus workers in other departments he should make the final decision himself.[81]

Where does this leave scientific research organisations, some of which are almost entirely ideological in character, but in all of which ideology is an important question? As far as we can tell from limited data, the essential distinction is between organisations of the Academy of Sciences and branch organizations. This is apparently the situation at Central Committee level, where the Science Department, under the Central Committee secretary for ideological affairs, is responsible for academic institutes, while branch institutes are supervised by the branch departments of the Central Committee. The same principle seems to apply at lower levels.

Thus when we look at party secretaries, the ideology secretary seems to be responsible, not surprisingly, for general ideological matters in science, but also specifically responsible for all the activities of academic institutes. V. N. Iagodkin, while secretary for ideology of the Moscow *gorkom*, wrote a notorious article on the state of ideology in Moscow research institutes,[82] while Z. M. Kruglova, the secretary of ideology of the Leningrad *obkom*, attended a conference on the social sciences.[83] These are matters of purely ideological concern. But the discussions in the presidium of the Academy of Sciences on the work of the Academy's Bashkir filial and the Urals Science Centre were not concerned with ideological questions. Nevertheless T. I. Akhunzianov and L. N. Ponomarev, the ideological secretaries of the Bashkir and Sverdlovsk *obkomy*, attended.[84] It was also Ponomarev who represented the *obkom* in discussions with *Pravda* over delays in the construction of the Urals Science Centre.[85]

But when an article is written about R&D and the industrial application of research the author is usually the secretary responsible for industry. For example, when A. P. Dumachev was Leningrad *gorkom* secretary for industry he spoke to a *gorkom* plenum on new technology in research and industry, while V. F. Malov, second secretary of the Irkutsk *obkom* wrote on *vnedrenie* in 1973.[86]

In areas of very high concentrations of research institutes, the party first secretary will often have some sort of science background, although there is no pattern to its nature. (This tends to apply only to *raikomy*, since cities and *oblasti* will rarely have such high concentrations of research facilities.) Boris Chaplin, first secretary of the Cheremushki *raikom* in Moscow,[87] has already been described as a mining engineer. The son of a well-known *komsomol* activist, Nikolai Pavlovich Chaplin, who was purged by Stalin in 1937, he graduated from the Moscow Mining Institute in 1955 and then went to work in the mines of Kemerovo *oblast'*. In 1961, according to Aleksandr Nekrich after the rehabilitation of his father, he joined the party and in official biographies is described as immediately becoming involved in party work. He would appear therefore to have begun his party career within the Moscow Mining Institute – he gained a candidate of technical sciences degree there in 1963 and was appointed *dotsent* in 1964. In 1967 he won the State Prize for work done on hydraulic coal mining methods in this period. By 1965 he was second secretary of the Oktiabrsky *raikom* in Moscow. It was while in this position that he handled the difficult problem of the takeover by 'liberal' historians of the party committee of the Institute of History. His restrained diplomacy helped him retain the respect of the historians, and also apparently the approval of the authorities, since in 1969 he was promoted to first secretary of the Cheremushki *raikom*.[88]

Chaplin's successor in Cheremushki *raikom*, V. T. Polunin, is another case of a party first secretary with some sort of specialisation in science. He is a candidate of technical sciences who went from Cheremushki *raikom* to head the Science Department of the Moscow *gorkom*.[89] V. A. Protopopov, first secretary of Moscow's Lenin *raikom*, is an economist but apparently of the technical rather than political variety. He did his candidate's degree in the Department of Accounting and Analysis of the Economic Activity of Industrial Enterprises of Moscow University. He was also secretary of the university's party committee.[90] One assumes that the Vladimir Vasil'evich Shiriaev who was secretary of the party committee of the Central Research Institute for the Technology of Machine Construction of the Ministry of Heavy, Energy and Transport Machine Construction in 1966 is the same Vladimir Vasil'evich Shiriaev who became first secretary of Moscow's Proletarsky *raikom* in the 1970s. Two first secretaries of the Sovetsky *raikom* on Novosibirsk would appear to be ideologists by training. R. G. Ianovsky is a doctor of philosophical sciences, while R. Vasilevsky is a doctor of historical sciences. Ianovsky went on to be head of the Science Department of the Novosibirsk *obkom*, deputy head of the Science Department of the All-Union Central Committee, and rector of the Academy of Social Sciences. The present first secretary of the Leningrad *gorkom*, A. P. Dumachev, while not apparently ever having worked as a scientist, has written extensively on scientific topics. He previously worked in and then headed the Industry Department of the Leningrad *obkom*.[91] These cases suggest that it is possible to make some sort of party career by specialising in science. But as yet no one has clearly got into 'party generalist' work by such a route, meaning that the road to the top is probably still closed.

Most *obkomy*, particularly those in industrial areas, now appear to have Science Departments within their apparatus. In fact the full title is usually Department of Science and Educational Institutions or Higher Educational Institutions. Like their equivalent in the Central Committee apparatus they seem to be responsible for Academy institutes, the social sciences and education. The staff of these local Science Departments show the same sort of bias towards an ideological background as does that of the Central Committee department, while according to Philip Stewart they are under the supervision of the party secretary in charge of ideology.[92] Thus two past heads of the Science Department of the Novosibirsk *obkom*, R. G. Ianovsky and M. Chemodanov, have philosophy degrees and write on ideological subjects.[93] S. V. Sakharov, head of the Science Department of the Kamchatka *obkom*, is a candidate of historical sciences who writes books for the ideology series of the Soviet Union's political publishing house.[94] A. A. Dobrodomov, once head of the Science Department of the Moscow *gorkom*, later became director of the Moscow *oblast'* Institute of Party History, while Z. S. Mironchenkova, department head in the Leningrad *obkom*, became director of the Institute of Party History of that *oblast'*. A former head of the Science Department of the Moscow *gorkom*, V. I. Tishchenkov, went on later to become deputy chairman of the RSFSR State Committee for Vocational and Technical Education, indicating that education is his main area of interest, G. P. Aleksandrov, head of the Leningrad *gorkom* Science Department, became deputy Minister of Culture for the RSFSR, also indicating an ideological orientation.[95]

Such people are not necessarily well qualified for any serious involvement in R&D management. Nevertheless these are the people responsible for science management when Academy institutes are involved. Thus Chemodanov writes on the introduction of research results into production in Novosibirsk, while the head of the Science Department of the Sverdlovsk *obkom* is involved in setting up the Urals Science Centre.[96] L. T. Rassokhin, head of the Science Department of the Komi *obkom*, has twice attended Academy presidium meetings on the work of the Komi filial of the Academy.[97]

It appears that the branch departments, specifically the Industry Department, of the regional party committees are responsible for the supervision of non-Academy industrial research. Thus A. Solov'ev, the deputy head of the Industry Department of the Leningrad *obkom*, wrote in some detail on the organisation of research work,[98] while the department's head in 1971, V. I. Pimenov, wrote on the effectiveness of economic reform in industrial research institutes.[99] See also V. I. Leonov, head of the Leningrad *gorkom*'s Industry–Transport Department, writing on branch science and science–production associations, and G. Palarmarchuk, head of the Industry–Transport Department of the Voronezh *obkom*, on science–production relations.[100] It was the head of the Oil, Gas and Geology Department of the Tiumen' *obkom* who responded to criticism in *Pravda* of the poor support given a Siberian oil industry research institute.[101] The heads of the relevant branch departments of the L'vov *obkom* act as deputy heads of the various science–production associations in the Western Science Centre, while control groups within the Industry Department help in the supervision of the programmes.[102] The Industry Departments of the Leningrad *obkom* and *gorkom* are involved in the work of the Leningrad Science Centre.[103] Party meetings in the Leningrad Research Institute of Chemical Machine Construction are attended by instructors from the *raikom* Industry Department.[104]

Clearly there is considerable scope for confusion in these arrangements. Particularly with the growing involvement of Academy institutes in applied research, including direct contractual relations with production enterprises, there seems to be room for jurisdictional conflict. Even where the division of responsibilities is clear, one feels that the staff of the Science Departments might be too constrained by their ideological backgrounds to provide useful assistance to research institutes, while the staff of the Industry Departments are likely to see their functions too much in terms of immediate production concerns, which again militates against assistance for serious research.

Note that in the dispute in Leningrad in the mid-1960s on whether enterprises should be transferred to institutes or vice versa, the head of the *obkom* Industry Department wanted research laboratories to be returned to enterprises.[105] One should also bear in mind that the regional party organs do not have large staffs. In 1978 there were 335 instructors in all the party committees of the Orenburg *oblast'*, and in the *gorkomy* of the Odessa *oblast'* each instructor supervised on average 49 primary party organisations and 2500 party members.[106]

While there is still much work to be done on an analysis of the training, skills and interests of local party workers involved in the supervision of science, it seems probable that the party does not have the staff in its regional party apparatus to support the role it would like to play in science management. This particularly applies to non-metropolitan areas, where high concentrations of research institutes in single geographical units are less likely.

This might be a blessing in disguise for Soviet science, in that it forces the party apparatus to rely on the support of scientists if it is to play any role at all. Firstly, let us look at scientist representation in local party bodies, though necessarily briefly, through lack of information. We will begin with committees of the party organisations, the large bodies that are equivalent at the local level to the Central Committee. Jerry Hough's data show that scientist membership is not at a high level. In 1966 three scientists were full members of the Krasnodar *obkom* (of a total of 117 members); two in Rostov (of 130); three in Ul'ianovsk (of 100); one in Volgograd (of 145); and none in Iaroslavl' (with 99 members). In the Rostov *obkom* the scientists were the director of the Don Agricultural Research Institute and the chief designer of a special design bureau. Hough also examined the Ul'ianovsk *gorkom* for 1966, which included two scientists as full members – the head of a design institute and a researcher at the Lenin museum.[107]

I have not had the opportunity to separate all the scientists out of other regional party committees, but a quick glance through the Moscow *obkom* and *gorkom* of 1980 reveals a few interesting names. In the Moscow *obkom* we find the missile designer Chelomei, the space engineer Glushko, and G. P. Svishchev, the rector of the Central Aerodynamics Institute. This is a remarkable collection of military research talent. The Moscow *gorkom* contains a couple of scientists less obviously connected with defence – V. A. Krasil'nikov, laboratory head at the Institute of Acoustics, and E. F. Oleinik, a polymer scientist at the Institute of Chemical Physics. Academicians on the *gorkom* include A. A. Logunov, the physicist vice-president of the Academy; G. A. Nikolaev, the welding expert who is rector of Moscow Higher Technical College; and M. A. Prokof'ev, the chemist who until recently was Minister of Education. The defence research establishment is again well represented, by G. V. Novozhilov, the General Designer at the Ministry of Aviation; Tupolev, the famous aircraft designer; and Piliugin, the rocket and space designer. The military presence is impressive, and corresponds with a similar strong presence in the Central Committee of defence scientists.

Not surprisingly, in areas with high concentrations of research institutions, and therefore mostly at *raikom* level, the representation of scientists is higher. Over half the membership of the Sovetsky *raikom* of Novosibirsk was made up of research workers in 1976. These included academicians Aganbegian, S. T. Beliaev, Marchuk and S. L. Sobelev, five corresponding members of the Academy, 12 doctors of science and 15 candidates.[108] In 1980 the Cheremushki *raikom* in Moscow appeared to be made up of roughly 30 per cent scientists, more or less equivalent to their representation among the *raion*'s party members. Eight full or corresponding members of the Academy and 28 doctors and candidates of sciences were included.[109]

The party committees are not of course decision-making bodies and are probably hardly even consultative. The membership of these bodies is almost purely symbolic. Membership of a regional party buro is, however, a different matter. The party buros are important policy-making bodies at the local level. Again information is scanty, but scientist membership of the buros appears to be rare. None of the *obkom* buros examined by Hough contain scientists, but he did identify the director of a research institute who was a member of the Erevan *gorkom* buro.[110] Academician N. A. Shilo, chairman of the Far East Science Centre, is a member of the buro of the Primor'e *kraikom* (and was once a member of the Magadan *obkom*). Krasil'nikov, the laboratory head of the Institute of Acoustics just mentioned, is – strangely, considering his apparent relative insignificance as a scientist – a candidate member of the Moscow *gorkom* buro. G. S. Migirenko, once secretary of the party committee of the Siberian Academy, and in the early 1970s deputy director of the Institute of Hydrodynamics, was in 1971 a member of the buro of the Novosibirsk *gorkom*. B. E. Bykhovsky was director of the Academy's Institute of Zoology, academician secretary of the Department of General Biology and a member of the Academy's presidium. He was also a member of the buro of the Vasil'eostrovsky *raikom* in Leningrad.

V. P. Mamaev, director of the Novosibirsk Institute of Organic Chemistry, is a member of the buro of the Sovetsky *raikom* buro. It seems probable that buros of other heavily scientific *raikomy* contain scientists, but without more evidence we cannot come to any firm conclusions on the extent to which scientists play a policy-making or influencing role through direct membership of party organs.

Scientists do have another opportunity to involve themselves in the work of the party apparatus. All regional party organs with an involvement in science management make use of the so-called party *aktiv*, that is, party members who although not full-time party workers are forced or prepared to devote much of their spare time to party work. In *raikomy*, with very limited full-time staff, these people are used in part-time Science Departments. In Moscow the part-time departments were set up in the mid-1960s. In 1965 the department in the Kirov *raikom* of Moscow had 32 staff, including 13 candidates and 2 doctors of science, organised into 6 sectors for different branches of the economy, plus sectors for design and drafting organisations. The department sends its staff into all the institutes in the *raion*, seeking possible improvements in research and work methods and examining plans in order to weed out out-of-date and duplicated research.[111] The Radiansky *raikom* in Kiev set up a non-staff Science Department in 1971. Its head is a candidate of technical sciences, while his deputy is a corresponding member of the Ukrainian Academy of Sciences. Most members are heads of departments, laboratories or chairs (*kafedry*). The department spends a lot of time working with primary party organisations, and prepares discussion papers for meetings of the *raikom* buro.[112] It seems that the part-time departments fulfil very similar functions to those of a full-time department.

Obkomy and major *gorkomy*, despite having full-time Science Departments, use activists in a whole range of councils and commissions. The most common councils are those for scientific and technical progress, for the implementation of research results into production, and technical–economic councils. To some extent they are used to tap the expertise of local scientists to help solve problems arising in local enterprises, but they are also used for the management and supervision of research. Thus the council of the Kuibyshev *gorkom* in 1977 had 105 members, headed by a presidium of 13 under the chairmanship of the *gorkom*'s second secretary. The council was divided into sectors, for scientific and technical information, social planning, the organisation of production and introduction of automated methods of management into production, the use of progressive technology in subsidiary production facilities, etc. The council meets four times a year, and at its meetings local industrial managers speak on the use of new technology in their enterprises. The council works out recommendations for improving R&D work in the city and on spreading the experience of leading enterprises. The recommendations are confirmed and put into effect by the *gorkom* buro.[113] The council of the Donetsk *obkom*, also headed by an *obkom* secretary, controls the fulfilment of research, design and drafting plans and strives to raise the effectiveness of the work of researchers.[114] In the capital cities the *gorkom* often sets up councils together with the presidium of the republican Academy of Sciences.[115] Some local party organs have councils for specific areas of concern. Thus the Kiev *gorkom* has 15 inter-branch commissions and 5 individual branch commissions, to which over 500 people are coopted. As an example, one of these commissions is charged with increasing the rate of technological progress in welding. It includes representatives of the Paton Institute, the All-Union Welding Design Institute of the Ministry of Lathe and Instrument Industry, the Faculty of Welding of the Kiev Polytechnical Institute, and a number of the city's industrial enterprises. In 1977 the commission supervised work on 97 contracts, of which 43 were 'contracts of cooperation'.[116] In the Ukraine at least, the work of the regional councils is controlled at republican level. The council for scientific–technical progress of the Ukrainian Central Committee (which is headed by party first secretary Shcherbitsky) heard reports from the first secretaries of the Dnepropetrovsk and Kharkov *obkomy* on the work of their *oblast'* councils.[117]

Such commissions and councils, bringing together researchers and production personnel, often of high calibre, could clearly play a useful role in promoting local technological progress. However, to the extent that they, and the part-time Science Departments, are to act as a substitute for a party organ's own specialist science apparatus, there are clearly problems. Firstly, if the party organ's task is to control and supervise the work of the research institutes, then using part-time departments and councils is in fact asking professional researchers to control and supervise themselves. Just as institute party secretaries will be shown in the next chapter to be likely to represent professional rather than party interests, so then are the part-time organisations. Secondly, if the task of the party organ is to help institutes overcome the difficulties and barriers they face in their work, then using the part-time organisations is in fact asking the researchers to help themselves. While they might provide a good forum for defining problems and bringing them to the attention of party workers, if those party workers do not have the expertise or will to take the matter further themselves, the influence of the part-timers could be limited.

The increase in the involvement of local party organisations in the management of science since the late 1960s is evident. This is presumably the result above all of Brezhnev's commitment to the scientific–technical revolution on the one hand, and to an increased role for the party apparatus throughout the Soviet system on the other. We have seen that the increased involvement has occured at two levels – in day-to-day control of and assistance to research scientists and their institutes, and in striving for the successful implementation of new approaches to the organisation of R&D.

It has not been possible, however, to demonstrate the effectiveness or general applicability of the two roles. At the day-to-day level the party organisations seem to have enjoyed a conspicuous lack of success in providing research institutes with the facilities they require. Lack of expertise and, probably, will would seem to be the major reasons, plus a probable lack of power vis-à-vis the enterprises of powerful ministries.

At the higher level of R&D reform, it has been possible to find evidence of an alliance between the party and reform-minded research organisations, particularly the Academy of Sciences, against the entrenched anti-innovation interests of the branch ministries. Such an alliance would suggest a significant and perhaps increasing role for L'vov-type programmes, giving as they do the opportunity for local party organs and the Academy to cooperate in the coordination of R&D programmes.

However one must retain a healthy respect for the inertia of Soviet bureaucratic procedures and the strength of the entrenched interests of the branch ministries. One would want to see more evidence of these factors being overcome before making any claims that the regional party organs are making an important contribution to progressive change in Soviet R&D.

4 Primary Party Organisations

The primary party organisation (PPO) has been described by one Western analyst as the CPSU's 'ultimate device of control and management.'[1] It is the 'ultimate device' because it is situated at the workplace level, at the centre of the working life of every member of the population, and also because it has extensive powers of control over both management and the rank-and-file working population. The areas of its operation tend to be described in extensive unbounded terms, with no aspect of the working collective's activity being closed to it.

The PPO becomes particularly important in the field of science and technology if our findings of the previous chapters are correct, that the professional party apparatus has somewhat limited interest and expertise in the science field. Being on the spot, research PPOs should not suffer from these difficulties, and the party might find itself with another opportunity to enforce the fulfilment of its science policies.

In this chapter I will firstly describe the functions of the research PPO and the means available to fulfil them, and then discuss its success or lack of success. Functions are divided into two main categories – ideological and the control of research.

IDEOLOGICAL FUNCTIONS

Given, firstly, that scientists deal professionally with concepts and phenomena that often strike at the heart of the fundamentals of Marxism–Leninism, and secondly, that they have on occasions demonstrated an above average willingness to hold unorthodox political views in their non-professional lives, it is to be expected that ideology would play an important role in the activities of research PPOs. Certainly it is true that during the worst excesses of Lysenko-

124

type interference in science institute party organisations were often the forum for attacks on and measures against targeted scientists.[2] It is also true that PPOs were expected to take a leading role against *podpisanty* and other dissidents in the late 1960s.[3]

It might appear that circumstances have changed substantially and that the PPOs' ideological role might now be greatly reduced. Lysenkoist methods are no longer permitted and dialectical materialism is broadly enough interpreted to allow acceptance of most mainstream and even some way-out theories, while a relative political calm has reigned in Soviet research institutes over the last decade. However while it is certainly true that some of the drama has been removed from PPOs' ideological activities, they are still taken very seriously. PPOs are still held responsible for what one can call in general terms the political health of the institute. This means that the PPO has to achieve a number of things. Firstly, all staff members should be politically and ideologically knowledgeable; secondly, that knowledge should be put into effect, in the sense of demonstrating a practical commitment to Soviet policy and ideology. This means most obviously an absence of dissent, but also requires more positive demonstrations of loyalty. Finally, political health means a healthy moral and social atmosphere, free of conflict and corruption.

What one might call routine propaganda and agitation continue, in order to educate institute staff politically and ideologically. PPOs do not have an enormous role in propaganda, in the classic Marxist–Leninist sense, since most of that basic socialisation is provided by professionals outside the work place, in the education and mass media systems. Nevertheless PPOs are expected to ensure that all members of the working collective have a basic grounding in Marxism–Leninism and are able when necessary to pass the information on to others. The institutes send their staff to the Universities of Marxism–Leninism organised by regional party bodies (and have great trouble getting them to see their courses through to the end),[4] or organise their own classes in the fundamentals of ideology.[5]

PPOs have a greater role in agitation. By agitation is meant the explanation of current and specific events in terms of the basic ideology of the Soviet Union. Events in international and internal politics are explained, and the tasks of particular organisations and even individuals in fulfilling national and regional policies are revealed. A whole network of lecturers and *politinformatory*, both professional and part-time, are used by PPOs to speak on such topics as 'Ideological diversions of modern imperialism', 'The scientific–technical revolution and the ideological maneouvres of bourgeois propaganda'.[6] Such activities are likely to be stepped up at times of political crisis, for example after the Soviet intervention in Afghanistan. On a less political note lectures, small-group conversations (*besedy*) and particularly party meetings, both open and closed, will be used to publicise the decisions of Party Congresses, Central Committee plenums, etc. The particular tasks of the relevant branch of the economy, institution or small group will usually be explained. 'Visual agitation' will also be used, with the PPO secretary supervising the hanging of posters, placards, etc.[7]

One of the most important features of ideological life in a Soviet research institute are the so-called philosophical (methodological) seminars. They would seem to have primarily an educative function, although over their longish history they have combined in varying proportions a number of aspects of the ideological function. Their basic function is as a forum for the discussion of links between Marxist–Leninist philosophy and individual scientific disciplines. There is some evidence that they are or have been used to lay down the 'correct scientific line' on particular issues. For example, at a philosophical seminar in the Latvian Institute of Physics in 1960 the participants were examining some articles that had appeared in *Voprosy filosofii*. In the course of the discussion a dispute arose over the commonality of matter and the possibility of establishing a single theory of fields, both gravitational and electromagnetic. The conclusion was reached that evidently the two types of field are of the same nature. The tone of the conclusion was that different views would not be tolerated.[8] In early 1964 it was claimed that seminars were being used to counter the influence of bourgeois idealists in Soviet science.[9]

But there is evidence that in the mid-1960s they were used, in a more or less official way, to restore respectability to the relationship between dialectical materialism and the natural sciences following the final disgrace of Lysenko. It was stated in *Vestnik Akademii nauk* in 1965 that the seminars prevented the formation of 'scientific monopolies' by allowing open discussion of philosophical issues.[10] In 1980 it was claimed that in the seminars 'in the relatively recent past basic attention was devoted to questions of the dialectical materialist interpretation of already developed theories of modern science (such as quantum theory, the theory of relativity, classical genetics, information theory, statistics and others).'[11] All the disciplines mentioned were in difficulty under Stalin and Khrushchev.[12]

However in the dissident atmosphere of the mid- to late-1960s things went too far, with even *samizdat* writers being invited to speak at seminars.[13] With the crackdown on dissidence in 1967 the seminars were brought under tighter control. Thus, for example, the Sovetsky *raikom* took over the management of the seminars in Akademgorodok, setting the topics for discussion and giving recommendations on how they should be run.[14]

In the early 1970s once order had been reestablished another effort was made to depoliticise the seminars. They were apparently taken from the formal control of the party organs and subordinated to institute management and the academic councils. They were further assimilated into the regular work of institutes by having the plans drawn up for them included in institutes' research plans. Standardisation was achieved by setting up a centralised control organisation under the Academy of Sciences.[15] This standardisation was designed to change the focus of their concerns somewhat. An effort was made to move from a narrow concern with the problems of a specific discipline to the large and perhaps less controversial questions of the methodology of science as a whole.[16] While philosophical and political themes are still evident,[17] a strong *naukovedenie* element is also appearing.[18] Certainly the apparent change in approach has been accompanied by an enormous increase in their number and the people said to be participating in them. Between 1970 and 1980 the number of seminars in Academy institutes almost doubled from 3700, with over 200 000 specialists involved in 1980. In 1974 there were 564 seminars in Moscow research institutes and *vuzy*, involving over 15000 scientists; by 1980 this had increased to 1134 seminars with 30 000 participants.[19]

Continuing complaints from some that the seminars are too little concerned with philosophy and ideology suggest continuing disputes over their proper functions.[20] One suspects that if scientists are to be involved in the seminars in any meaningful way (the other regular complaint is the apathy with which they are regarded), they will try hard to maximise their 'research' orientation. Higher party authorities could well be prepared to go along with that if it removes the possibility of the scientists using the seminars again to discuss 'dangerous' political topics.

It is of course the task of the PPO to give the seminars the correct balance between ideological and research concerns. Despite the fact that the seminars have apparently been removed from direct formal subordination to the party organisation, it should not be thought that the PPO, or indeed the *raikom*, has no interest or involvement in them. Reports can still be found, as one would entirely expect, of party buros examining the work and findings of seminars.[21] It was declared in 1979 that each seminar has a bureau, usually headed by the institute director and including as members representatives of the party buro.[22] In 1980 I. A. Glebov, the head of the Academy of Sciences in Leningrad, commented on the importance of party assistance in running the seminars.[23] One of the latest references refers to general leadership by the party committee and its ideological commission, the *partkabinet* and methodological council.[24] Ianovsky declares that PPOs must closely examine programmes and work with seminar leaders in order to ensure that they do not lose their philosophical content,[25] while the party committee of the Academy's Institute of Chemical Physics examines seminars not less than three times yearly to check that they are not becoming too narrowly professional.[26] Their success or lack of success in this concern will be part of our general evaluation of the work of PPOs.

Considerable scepticism, which is reinforced by regular complaints in the Soviet press, would be due as to the enthusiasm or seriousness with which any propaganda or agitation activities are regarded by Soviet scientific personnel.[27] Nevertheless they are only a small part of the PPO's responsibility (shared with management) for the political health of the institute. That requires more than an absence of dissent, or even passive acceptance of Soviet ideology and policy, but rather a positive and publicly-acclaimed commitment by the collective to all party policies. Scientists are regularly and deliberately put in a position of having to make such acclamations. In recent years the most moral-sapping demands of loyalty have been over the 1968 invasion of Czechoslovakia and during the campaigns against Sakharov and Solzhenitsyn. One is put in the position of having to vote for a resolution supporting Soviet intervention or sign a letter condemning Sakharov, or lose one's job. Non-party members can sometimes, although by no means always, avoid such demands, which fact alone leads many to avoid party membership. The determination with which the authorities pursue such declarations of loyalty suggest that they at least feel that obtaining them increases loyalty, even if only by breaking down feelings of independence and moral worth.

Political health means more than maintaining a proper degree of public loyalty to Soviet ideals. It also often means taking account of more specific political situations, in particular the factional line-up within the political leadership or the policy interests of various leaders. Thus the PPO of the Institute of Economics was expected to pick the right side in the debate over 'A' and 'B' sector investment in the mid-1950s,[28] while it was the party buro of the Shortandy Research Institute of Grain Growing that took measures against the institute's director Baraev for persisting with work on clean fallow methods despite the insistence of Khrushchev that such methods not be used.[29] In the social sciences such political awareness and flexibility can still be important, as the cases of sociology and particularly economics, illustrate.[30]

Political health also means that the institute be free of illegal or corrupt behaviour on the part of management and of unseemly personal or professional conflicts. Control of such things if formally placed by the Soviet authorities in the ideology category for two reasons. Firstly, such phenomena, particularly corruption, drastically damage the legitimacy of the whole regime in the eyes of the people. Secondly, difficulties in this area are claimed to be the result of failures in socialisation and educational work.[31] The extent of corruption and conflict in Soviet life leaves the PPO with plenty of work to do.

To sum up, undoubtedly the most important ideological function of the PPO is preventing the appearance in an institute of, worst of all, anti-Soviet or anti-party views. However maintaining the political health of the institute entails far more than this. The more formal principles of Marxist–Leninist ideology also have to be seen to be observed. Also the institute should be characterized by moral cleanliness, with an absence of corruption and personal and work-related conflict. Finally, and probably least wholeheartedly, the entire collective should approach their professional and social duties with enthusiasm and genuine commitment.

CONTROL OF RESEARCH

We speak here of the PPO's control of the actual research being done in institutes, including control of planning, assisting in the research process and verifying fulfilment. This entails keeping an eye on both management and the rank-and-file. In normal circumstances the emphasis should be on cooperation with management, control becoming little more than working with management to get people to work harder and better. However the potential of the PPO in unusual circumstances, when decisive action is needed against management, cannot be ignored.

In an earlier article I have provided detailed information on the work of PPOs in these areas.[32] I suggested there that PPOs play a considerable role in the planning process, in ensuring that institute plans are relevant to national priorities, that small-scale projects are eliminated, and that only essential adjustments are made during the plan implementation period. In recent times they have become involved in controlling the implementation of the new goal-oriented planning systems.[33] I gave some examples of PPOs becoming involved in solving particular problems in research projects, by encouraging contacts with scientists from other sections of the institute or from outside.[34] At times they come close to playing an 'expediting' role.[35] I also found that the greatest emphasis on increasing the role of research PPOs was at times when there was a parallel emphasis on *vnedrenie* (as reflected in the demands on the Academy in this area). This suggests that this task is seen as one of their most important areas of work. As part of such a task great emphasis is put on their responsibility for the control of contracts with production enterprises.

I also described in the earlier article personnel management as one of the PPOs most important tasks. Personnel control is of course both a task and a means to fulfil other tasks. As mentioned above, party control of personnel became a major factor in the second half of the 1960s, as part of the ideological crackdown of the time. Particularly once the 'liberal' party buros and committees that were reasonably frequent in the mid-1960s were removed,[36] the PPO became a major weapon of the party in its efforts to cleanse Soviet research institutes of ideologically unreliable individuals. However the emphasis on PPOs' personnel powers has not slackened since the apparent lessening of ideological tensions, with it being made very clear since the early 1970s in particular that PPOs have the right to examine all personnel movements.[37] There has been some dispute over whether PPOs need or are allowed to have formal *nomenklatury*. The official view seems to be that, since the PPO oversees all personnel movements, a *nomenklatura* is not necessary.[38] However, apparently PPOs, particularly in large institutions, in practice concern themselves only with more senior positions and for these positions they operate very much like higher level party organs.[39]

The suggestion now is that the PPO is expected to concern itself with more than the political reliability of personnel, but also with their scientific capabilities. Ostensibly in order to improve the quality of research personnel PPOs are given the right to make recommendations for promotion. A Soviet source declares that since they are in a good position to judge the capabilities of workers down to the lowest levels, PPOs should take the initiative in suggesting people for promotion and even maintain 'reserves for promotion'.[40] Sometimes the PPO's recommendation will be for some party hack looking for a peaceful spot to retire,[41] but we can presume that party organs, usually made up of leading members of the institute, also have some genuine interest in raising the scientific quality of personnel.

The same motivation presumably has some relevance to the various attestation, material incentive and socialist competition schemes that have received so much attention since the late 1960s.[42] The PPO is directly involved in these schemes, with it being expected to exert its influence on the formation of all commissions, including placing its representatives on any such bodies and ensuring that they vote according to the party line. The party secretary signs all resulting recommendations and decisions.

The requirement that the PPO provide a *kharakteristika* (reference) for anyone wanting to move jobs is also ostensibly to help ensure the quality of Soviet research staff. Whether the researcher is a party member or not, such a reference is necessary for any transfer or promotion and, incidentally, any trip abroad.

MEANS OF CONTROL

The treatment of the PPOs' personnel function could easily have been included as one of the means available to PPOs. Their involvement in personnel management is important in maintaining the political health of the institute, including when non-party members are involved. All the personnel powers just mentioned are equally applicable to party and non-party members. They were used to great effect, although as we will see with some reluctance, in the crackdown of the late 1960s. Since then the opportunities to use such methods have become even greater. The opportunities for abuse of attestation and material incentive schemes and the *kharakteristika* are obvious, and even openly feared by some Soviet writers.[43] They can be used against both politically undesirable researchers and incompetent or dishonest managers.

In most cases, particularly the latter, such measures would be used together with party disciplinary measures. The PPO has the right to take direct party disciplinary measures against any member. Lesser penalties include the various grades of reprimand (*vygovor*): warning (*postanovka na vid*), ordinary reprimand, strict reprimand, either on the party card or not. There are claims that managers collect such reprimands as a matter of course and do not necessarily take much notice.[44] However the ultimate penalty is expulsion from the party.[45]

This will almost certainly be accompanied by dismissal from one's post and indeed removal from the entire *nomenklatura* 'carousel'. Party membership among scientists is high, and is very high among management personnel.[46] Therefore the PPO's leverage is considerable.

In cases of political misdemeanours or illegal and corrupt behaviour party discipline will probably be applied immediately. In the PPOs' more routine control of the scientific process it is more likely to be used as a last resort, after the PPOs' managerial powers have failed to produce the desired effect. These powers are summed up in the right of control.

The right of control is a somewhat difficult concept, which has received rather inconsistent treatment in the Soviet press. It made its first appearance at the 18th Party Congress in 1939, when it was limited to the PPOs of production organisations. It was stressed at the time that PPOs were in fact already using the powers of the right of control, but that the new right formalised and made more secure those powers.[47] The same sorts of statements were made in 1971, when the right of control was extended to the PPOs of non-production institutions, including research institutes.[48] Particularly in the case of research institutes the granting of the right of control was largely the culmination of an increasing role for the PPOs since about 1967. Nevertheless there is a general consensus in the Soviet press that the 1971 right of control did have an important effect on the activities of non-production PPOs, by increasing the degree to which the PPO could reliably hold management accountable, making its control activities less episodic, and to some extent expanding its interests more determinedly into the control of research and management, rather than purely ideological questions.[49]

The situation for research PPOs was complicated by the fact that some research PPOs had been given the right of control ten years earlier at the 22nd Party Congress. Even before that, in June 1959, the PPOs of 'drafting offices, design bureaus and research institutes fulfilling orders (*zakazy*) for enterprises and construction projects' suddenly appeared in the Regulations (*Polozhenie*) attached to the Central Committee decree 'On the formation in the primary party organisations of production and trade enterprises of commissions for the implementation by party organisations of the right of control of the activity of the administration of enterprises', as being among those PPOs having the right to establish such commissions.[50] This is strange, firstly, because they were not mentioned in the decree itself, and secondly and most strangely, because they did not have the right of control. They were being allowed to establish commissions to implement a right they did not have. Given the ambiguity of the commissions' existence, it is not surprising that after an early burst of enthusiasm little was heard of them.[51]

To some extent the ambiguity was removed at the 22nd Party Congress, when the PPOs of 'drafting offices, design bureaus and research institutes directly linked with production' were granted the right of control.[52] However considerable ambiguity still existed in the definition of what constituted a 'direct link with production'. The only article to make any attempt to explain the production/non-production distinction spoke of institutes of an industrial profile and those of a 'so-called academic (theoretical) profile', with the latter usually being Academy institutes.[53] We have little information on which institutes found themselves in which category, since there was surprisingly little coverage of the change or of research PPOs in general in the next few years, presumably a consequence of the confusion and complete absorption of the party apparatus in short-term production concerns, diverting their attention from R&D matters. But there are indications that the 'link with production' clause was interpreted narrowly. Thus the drafting bureau '*Leninskaia kuznitsa*' had the right of control, but the Research and Experimental Design Institute for Automisation of the Non-Ferrous Metallurgy Industry did not. Ianovsky includes the experimental factory of the Siberian Academy of Sciences in a list of institutions that put the right of control into effect after the 24th Party Congress.[54] Thus it seems that experimental work, at least within the Academy of Sciences, was not considered to be sufficiently closely linked with production. On the rare occasions when reference was made to the right of control in production-oriented institutes, it was stressed that PPOs without the right of control were also expected to involve themselves in the work of their institutes.[55] In recent times the 1961 rule is largely ignored, as if all research PPOs received the right of control in 1971.[56]

It is certainly true that after 1971 all research PPOs were on an equal footing. What then is entailed in their powers? The powers of the right of control are summarized as follows:

1 to hear reports from management when necessary
2 to establish permanent and temporary commissions
3 to study affairs on the spot and to be acquainted with relevant information

4 to offer suggestions and recommendations, which must be taken into consideration by management, and to strive for their implementation.[57]

Point 1, the right to hear reports from managers when necessary, appears in practice usually not to apply to institute directors. Authoritative statements stress that any leading official is obliged to give reports before the PPO,[58] and yet lists of research managers that actually do so usually exclude the director.[59]

The other feature of the coverage given to the hearing of reports in the Soviet press is that it usually refers to the hearing of reports from 'communist leaders', despite the fact that authoritative statements make it clear that the party status of the manager involved is irrelevant.[60] Most leaders of Soviet institutions are party members and are therefore bound by party discipline to account for their behaviour before their PPO. Presumably this is an indication of some controversy over the hearing of reports. Party officials prefer to avoid any possible difficulties by reference to the requirements of party discipline.

Another important matter to be noted is a difference in language that disappears in translation. There are three Russian words that all translate as 'report'. *Soobshchenie* and *doklad* tend to have the sense of reports primarily for information purposes. The speaker simply reports on the state of affairs in his section of the institution or even on some very general matter. These words do not imply accountability. However the word *otchet* has the sense of accounting for one's behaviour. It would appear that an *otchet* is a more serious matter. A case could be made, based on empirical data, that administrators who are not party members, or who are not appearing as party members, are more likely to give *doklady*. That is, administrators try to avoid giving *otchety* unless party discipline forces it on them. For example, in 1971 the secretary of the party committee of GKNT wrote that meetings of the party organisations, party buros and party committee of the State Committee, in order to influence the work of all its sub-divisions, hear *doklady* from sub-division leaders and *otchety* from communists.[61]

However this linguistic difference does not fully explain the confusion in Soviet commentaries and practice. The list of powers entailed in the right of control given above uses the word *otchet* and makes no mention of communist leaders, and some specific examples of such *otchety* being heard from non-party managers can be found.[62]

Although the right of control, strictly speaking, refers to hearing reports from institutional leaders, particularly in recent years there has been considerable emphasis on the desirability of PPOs hearing reports from rank-and-file workers on the fulfilment of their official duties. Usually the people giving such reports are party members subject to party discipline.

It appears that, formally speaking, all PPOs have always had the right to hear reports of all kinds from anyone. In practice, reports, particularly those implying accountability, are usually limited to those leaders below the top level of the institution and spoken of in terms of a report from a party member as much as a manager. The extent to which reports are heard and the degree of their significance depends primarily on the general policy on such practices established by the top authorities at the time. At the moment they are an established part of the PPOs' administrative control.

The second element of the right of control listed is the right to establish permanent and temporary commissions. The most formally defined of these are the permanent commissions for the exercise of the right of control, established by the Central Committee decree of June 1959 mentioned above. A recent decree redefined and reaffirmed the tasks of the commissions.[63] Both decrees declared that there was a need for systematic control of the prompt completion by enterprises of their tasks. The commissions were specifically designed to provide an institutional means of control, to end the practice of party control being limited to hearing reports from managers on plan fulfilment. It was stressed that the commissions were not to work parallel to the PPOs, but were to work within them and under the control of the party buro or committee and its secretary. Throughout the period of their existence it has been stressed that the commissions are not to be used for general control of all the activities of institutions, but are to be used for examining particular important areas. The sorts of commissions in research PPOs are for scientific research, for the development of the scientific–production base of the institute, and for *vnedrenie*; or in another institute, for speeding scientific progress, for raising the effectiveness and quality of research and design work, and for progress in construction work.[64]

Commission members are elected by open vote at general party meetings, although changes can be made by the party buro or committee. Commissions do not take decisions themselves, but send their recommendations to the party committee or sometimes direct to a party meeting.[65] The chairman of a commission is usually a deputy

secretary of the PPO.[66] The opinion is usually expressed that there is
no need for permanent commissions in small PPOs, since in them all
necessary control can be exercised by the party organisation as a
whole.[67]

PPOs also make considerable use of temporary and special-purpose
party commissions. It appears that the main function of these commissions is obtaining on-the-spot information and controlling the fulfilment of very specific tasks. For example, the commission set up in
the Minsk Research Institute of Computing for the introduction of a
faultless work system was disbanded once the system was operating
properly.[68]

The PPO also uses non-party commissions to collect information.
The 'people's control' system is a good example.[69] Like other social
(*obshchestvennye*) organisations, 'control groups' enable PPOs to use
non-party members in the exercise of their functions. The primary
task of control groups is information-gathering and in this they are
very like the party commissions just described. They can be established
in all collectives, production and non-production, and they have the
right to examine documents and information, to receive figures,
reports and explanations from the director and his subordinates, and
to issue directions (*ukazaniia*) on the removal of shortcomings.[70]
Although it would appear that some of these powers are rarely used
fully, they are clearly important powers that could be very useful to
a PPO.[71]

Some idea of the activities of control groups can be gained from
a report on the activity of the group in the institute *Gidroaliuminiia*
in 1963.[72] The group checked how the most recent technical achievements were being used in new designs and equipment. It uncovered
many shortcomings in designs used in enterprises concerned with
electrical production. The chairman of the group sent the results of
the investigation to the party organisation of the institute. The director
and the party buro then had a conference with department heads and
chief specialists to work out what to do. This appears to be typical of
the groups' activities. They conduct an investigation and find shortcomings which they report to the party organisation which then takes the
necessary action.

The third element of the right of control, the right 'to study affairs
on the spot and acquaint oneself with relevant information', is a
summary of the whole information element of the right of control,
of which points 1 and 2 are the administrative expression. However
its inclusion as a separate point indicates that formal hearing of reports

and the establishment of commissions are not the only ways of obtaining information.

Although I do not have any specific statements to the effect that managers are obliged to furnish PPOs with all draft decisions on a routine basis, there are sufficient reports of this happening in practice to suggest that it is an informal rule. It appears to apply particularly in personnel and planning matters.[73] However statements to the effect that 'nothing is done in the collective without the approval of the PPO' suggest that it applies to all areas.[74]

Another important source of information is informal contacts between party officials, particularly the party secretary, and management. The Soviet press often mentions the need for party business to be conducted whenever suitable outside constraining and time-consuming meetings. An *obkom* instructor wrote in 1972 that the right of control raises the role of the PPO secretary in that people will more often come to him so that he, together with the director and other responsible people, might decide production, cultural and welfare (*bytovye*) matters in the course of work (*v rabochem poriadke*), without scheduling a meeting of the PPO or its committee or buro. He declares that 'this personal contact, this exchange of opinions, undoubtedly is a form of control and has its uses.'[75] However it is unlikely that management is under any formal obligation to engage in such conversations.

Finally we should not forget that rank-and-file party members are expected closely to observe the everyday work of members of the collective and report any possible shortcomings to the PPO.

Point 4, allowing the PPO to make suggestions and recommendations, introduces a fundamentally new element into the right of control. Points 1, 2, and 3 gave the PPO unlimited access to information. The only matter over which there were significant differences of opinion or practice in Soviet sources was reports by managers to party organs and organisations. However these differences occur as the result of the unwillingness of managers to be put in a position where they are forced to account for their behaviour before the PPO. To this extent the disagreement is more closely connected with the essence of point 4, that is, what powers does the PPO have to act on the information it receives and analyses. This is a complex and obviously sensitive question. It is a question the answers to which are often contradictory both in theory and practice, and one on which authoritative commentators are often loath to commit themselves. I will try to set out the different opinions I have come across, and to explain some of these differences, and then give a brief summary of what happens in practice.

Most lists of the powers contained in the right of control, such as the one I have given above, state that a manager must consider the recommendations made by the PPO (the usual phrase is *ne mozhet ne schitat'sia*).[76] But to simply consider a recommendation is not to do very much – the important thing is whether management is under any obligation, in theory or practice, to act on the recommendation.

The great majority of the cases appearing in the Soviet press assert that recommendations have in fact been accepted and acted upon, and many give the impression that management had little choice in the matter. However these reports are concerned with cases where the PPO has successfully dealt with the situation and where, as far as we can tell, management has simply accepted its recommendations. There could be many reasons for such acceptance, none of them necessarily implying any obligation on the part of management.

There is no formal obligation on management to accept recommendations, since that would constitute *podmena*. Thus *Partiinaia zhizn'* wrote in 1955:

> The primary party organisation does not have the right to change the orders of the administration or to give any type of orders concerning production work, to demand (*trebovat'*) that the director, the fully empowered master (*khoziain*) of the enterprise, submit his orders for the party organization's approval; nor may it remove, appoint or recommend workers in production. These are the functions of the administration,[77]

and in 1971:

> May a primary party organization, having the right of control, bring to account the leader of an institution? May it oblige him to do this or that? Is that what control means? Of course not.[78]

Nevertheless some Soviet reports do refer to the right of PPOs to give obligatory orders in exceptional circumstances.[79] Such a case appeared in *Partiinaia zhizn'* in 1978. By a decision of the party buro of the institute *Orgstankinprom* the chief engineer and deputy director for research 'were obliged (*obiazany*) to remove shortcomings [in the institute's financial situation] that had been uncovered in the course of the discussion, to restructure planning and to improve the quality of the fulfilment of orders.'[80] On occasions PPO directives are couched in very specific terms. Thus the PPO of the Institute of Social Sciences of the Ukrainian Academy of Sciences 'directed' the Department of Ukrainian Language to study the role of the Great Russian language as a means of international communication. The political implications of the directive are obvious; nevertheless the language used is unusual. A party meeting in the same institute 'obliged the directorate to make strong representations to the presidium of the Ukrainian Academy of Sciences that the admission of post-graduate students be broadened.'[81]

These last two cases seem to be quite contrary to the usual principle of PPO–management relations.[82] Even the more general directive given in the first case is unusual. It is interesting to see the criticism of a factory PPO for a very similarly worded decision in the same journal at almost precisely the same time.[83] However it does at least conform with the idea of control as not extending as far as giving direct orders on specific matters to management. Such a decision is not a specific order, but rather official notification to management that the PPO is dissatisfied with the state of affairs and that some remedial action is expected.

As we have already seen, the PPO can use its party disciplinary and personnel powers to apply or threaten sanctions against a recalcitrant manager. Before resorting to such measures it is most probable that the party secretary would turn to higher party authorities. This would particularly apply if the case involved the head of the institution. The system is deliberately organised so that higher authorities will immediately know if the state of affairs in any institution becomes serious. The consequences of error are so severe that no PPO secretary would be willing to risk a full-scale assault on an institutional leader without at least being strongly confident of support from higher party authorities. As we will see later the Soviet press is full of complaints from PPO secretaries who feel that they are not getting the necessary support from their *raikom* or *gorkom*.

Therefore means have been developed of attracting the attention of higher party bodies. One is failure to elect an institutional leader to the party committee or buro. This is probably not a common example of the use of party membership to influence the behaviour of an institutional leader, but it is one that is highly effective. It is unheard of for the head of an institution (assuming he is a party member) not to be a member of the party committee or buro for any period of time.[84] The organisation could not operate in anything like the proper way in such a situation and the *raikom* or *gorkom* would have to take quick action if such a situation arose. A case reported in *Partiinaia zhizn'* in 1955 illustrates this well. The PPO of a research institute failed to elect the director of the institute to the party buro. The *raikom* investigated and found serious shortcomings in his work. The *raikom* sent the information to the relevant ministry, upon which the collegium of the ministry dismissed him.[85] In 1963 the party members of the Moscow Design-Construction Office *Glavsantekhmontazh* of the State Production Committee for Assembly and Special Construction Work were dissatisfied with the head of the office. At the election-report meeting he was nominated for election as a member of the party buro. Obviously knowing what was to come, he refused the nomination. But the party members rejected his refusal, placed him on the ballot, and then failed to elect him. The consequences are not recorded.[86]

Another means of attracting the attention of higher authorities is to use party meetings to criticise institutional leaders when it is certain that the authorities will become aware of the proceedings of the meetings. For example, at the election-report meeting of the PPO of the Ukrainian Institute for the Design of Heavy Machine Construction Enterprises, the director and chief engineer were criticised for poorly organising and planning the production process. This was probably for the benefit of the *raikom* first secretary who was present at the meeting.[87]

The reliance on higher party bodies must play an enormous role in the informal relations between PPOs and management. But higher party bodies figure prominently even in the formal expressions of PPO powers. Many of the statements regarding laying party charges against managers add that this can be done only after application to higher party bodies.[88] This is therefore a formal as well as practical necessity.[89]

To summarize, PPOs have the right to make recommendations to management, recommendations that should carry some weight. It appears that in some circumstances they are able to oblige management to act, but even then they are not able themselves to make the decision on what is actually to be done. The directives can only be generally worded demands for action.

It is difficult to say how seriously management has to take these generally worded orders. The fact that Soviet sources stress that such orders can be given only in exceptional circumstances suggests that on those rare occasions a manager would do well to take notice of the PPO. A PPO driven to such extremes, if it had not already conferred with higher party authorities, would at least be confident of their support. In such cases some rapid improvement in the situation is expected and probably obtained.

The more usual case, when the PPO offers a recommendation, is even more uncertain. The vast majority of reports in the Soviet press declare that recommendations were accepted without disagreement. However cases where recommendations were rejected are perhaps less likely to be reported. A number of other factors also have to be mentioned. Firstly, there is a strong possibility that recommendations have already been agreed upon by the PPO and management. Senior managers are almost always members of the party buro or committee, the organs where final decisions, including the recommendations to be presented by general party meetings, are made. In normal circumstances any differences of opinion are presumably worked out in the meetings of these organs. Our knowledge of what happens in these meetings is extremely limited. Complaints about management using party meetings simply as forums for giving operational orders suggest that party recommendations could often be little more than party support for such orders.

At times when the role of PPOs in the control of research is being particularly emphasised, at which times the PPOs are more likely to receive support from higher party authorities, it is possible that recommendations are not arrived at in collusion with management. In such situations the general state of party–state relations and support for the PPO from higher authorities make it probable that recommendations will be accepted.

It can be seen that the powers of the right of control are of great potential and probably the most important means at the disposal of the PPO. One final important means remains to be mentioned, and that is the PPO's influence over a working collective's collegial bodies, in the case of a research institute the Academic Council (*ucheny sovet*). While the firm Soviet administrative principle of *edinonachalie* (one-man management) means that such bodies rarely have the ultimate power of decision or instruction, they do have considerable recommendatory powers. Thus the Council has extensive responsibilities in all aspects of institute life – it examines research plans of individual researchers and of the institute as a whole,[90] and considers appointments and reappointments.[91] Conflict between the director and the Council is not impossible. Nevertheless the Council would seem to be primarily a secondary instrument of management.[92]

The PPO has extensive opportunities to influence the operations of the Council, mainly through its control of the members who are party members. They appear always to make up a majority – often a bare majority – of the membership,[93] The PPO exercises discipline by operating a party group which is informed of the correct line before the meeting,[94] and hearing reports (*otchety*) from communist members of the council.[95] It will also use other less direct means, such as examining the activities of the council in party meetings or control commissions and making recommendations to management.[96] In the early 1970s, when PPOs had *nomenklatury*, council members were listed.[97]

The councils are used for political control, both in getting rid of politically unreliable researchers,[98] and condemning ideologically unsound research work,[99] as well as for more routine control of the research process.

It can be seen that the powers of the PPOs run broad and deep. They are expected to play a major role in all aspects of institute life, including such major areas as political and ideological control, personnel management, and control of the planning process. The means available are also impressive, particularly the actual or threatened application of party disciplinary measures and the formal powers of the right of control.

THE EFFECTIVENESS OF PPOs

Nevertheless in evaluating the actual powers and effectiveness of PPOs it is necessary to realise that the PPO is not expected and is not given the power to 'run' the institute. It has no power to give operational orders to management or the rank-and-file, while even excessive informal interference in the day-to-day running of the institute is likely to attract charges of *podmena*. The PPO exists essentially as a back-up to management in its efforts to get the best out of the collective, given that institute management will usually be carefully selected by higher authorities in order to ensure that it will be well-disposed towards party policies. As a Soviet handbook says:

> The right of control does not only not reduce the principle of one-man management, but on the contrary strengthens it. It makes possible a raising of the authority of the administration, strengthens its influence on all aspects of the activity of the collective, strengthens labour and production discipline, and improves the situation of all affairs in all areas of economic and cultural construction.[100]

However the PPO is also there as a check on management, in both a routine way, to ensure that party policies are being fulfilled, and in exceptional circumstances, when such policies are being deliberately frustrated or management is involved in illegal behaviour. The powers granted the PPO – party disciplinary powers, personnel control and the right of control – should be enough to ensure the successful fulfil- ment of these tasks.

However we are struck by the number of accounts of PPOs failing in their responsibilities. These failures have occured in all areas of their activities. Perhaps the most striking are in the field of ideological control. I will limit myself here to a few examples, from different periods and circumstances.

In 1937, as part of Lysenko's campaign against Nikolai Vavilov, a *Pravda* article criticised not only Vavilov, the director of the Institute of Plant Breeding, but also the institute's PPO secretary, for failing to prevent various destructive genetics-inspired practices and for har- bouring politically unreliable people in the institute.[101] In 1955 the secretary of the PPO of the Institute of Economics of the Academy of Sciences was dismissed from his post by the Moscow *gorkom* for refusing to admit that views expressed in the institute's Academic Council in support of the priority development of sector 'B' investment were anti-Marxist. The political significance of this dispute is obvi- ous.[102] In 1957 the Central Committee took action against the PPO of an unnamed institute for allowing a small group of scientists, under the pretext of criticising the cult of personality, to 'attack the policies of the party' in a party meeting.[103] In 1960 the party buro of the Latvian Institute of Economics was criticised for failing to deal with the political mistakes of the director. He was accused of having directed the institute's research toward the development in Latvia of only those branches of the economy the means of production of which were locally available, and having put undue emphasis on the con- sumer industry.[104] In 1966 Aleksandr Solzhenitsyn found that invitation after invitation to speak in research institutes was withdrawn following party pressure, but suddenly at the Institute of Oriental Studies he was permitted to speak by the institute's party committee, and indeed party committee members shared the platform with him.[105] Valery Pavlinchuk, one of the prime movers in the development of free social discussion in the Physics-Energy Institute in Obninsk, and eventually expelled from the institute for possessing anti-Soviet literature, was the party group organiser of the institute's Theoretical Department.[106] In 1969 a major campaign against the increasingly self-confident dis- cipline of sociology centred around Iu. A. Levada, the secretary of the PPO of *IKSI*. He was removed from his post after refusing to accept that his course of lectures on the theoretical basis of Soviet sociology was insufficiently grounded in historical materialism.[107] There is also the failure or reluctance of a number of PPOs and party committees to take action against *podpisanty* in the late 1960s.[108]

Cases of the punishment of institute management for illegal behaviour almost invariably include the party secretary in the list of the condemned, either for direct involvement or for passivity. The behaviour described is usually a matter of falsifying documents in order to receive unearned bonuses, illegal construction and allocation of flats, spending institute funds on banquets and parties, etc.[109] Sometimes the misdeeds are described vaguely as 'violations of personnel-financial discipline, poor management (*bezkhoziaistvennost'*) and squandering (*razbazarivanie*) of material assets.' In these cases there is no hint of corrupt practices, but simply of incompetence or of normal bending of the rules suddenly attracting attention. Again in such cases the PPO is invariably implicated.[110]

Criticisms of PPOs for failing to fulfil the role expected of them in the control of science management are numerous. Often they take the form of criticism for not making full use of the right of control or, before research PPOs had the right of control, for claiming that the lack of it limited them to insignificant matters.[111] Usually the criticisms reflect the general priorities in party control of research – personnel and productivity, ensuring research is conducted in fruitful areas and that research results are applied to production, and the general quality of research.[112]

One has to be careful generalising from critical reports in the Soviet press. No matter how often such reports appear they will cover only a small fraction of Soviet institutes, while there is probably a significant bias in Soviet reporting towards the critical, strengthened by the principle of criticism and self-criticism. However failures in ideological control strike so directly at the heart of the Soviet control system that even scattered reports are impressive, while the constant complaints in the press of continuing problems on a nationwide scale in science planning, labour productivity, research effectiveness, and poor *vnedrenie* suggest that PPOs in general are failing in these aspects of their activities. The feeling that the necessarily scattered criticisms of institute PPOs perhaps have systemic significance is strengthened by the fact that convincing reasons can be found to explain the problems. These reasons can be divided into two basic categories – that PPOs are unable to fulfil their functions, or that they are unwilling to do so. The first is concerned with the powers available to the PPO; the second with the motivations of PPO members, particularly the secretary.

The power of the PPO

I have already claimed that particularly since 1971, with all research PPOs having the right of control, PPO powers have been extensive and probably adequate. However some qualifications have to be made.

Firstly, there is the uncertainty that some PPO secretaries display about the extent of their powers. We will soon see that most research PPO secretaries are only part-time party officials. This, plus the deliberate ambiguity that is built into PPO–management relations, coupled with, historically at least, anomalies in the granting of the right of control, makes it unsurprising that some PPO secretaries exhibit a degree of uncertainty about the tasks for which they are responsible and the means available to fulfil them.[113]

The feeling of uncertainty could well be a factor in some of the failures of PPO secretaries to control the illegal behaviour of individual managers. It is recognised by the Soviet authorities that in some cases illegal administrative procedures must be tolerated, since legal procedures have shown themselves to be inadequate. It is part of the PPO secretary's role to decide when an action of dubious legality is justified by circumstances. There are some cases where it seems possible that a PPO secretary has been criticised for failing to take action over a matter that at other times might not have been considered as requiring any action. For example, the PPOs of the RSFSR Ministry of Consumer Services and the ministry's Central Design-Construction and Technical Bureau were critised in 1974 by the Committee for Party Control for failing to take action against the director of the bureau and the ministerial *glavk* supervising it after the bureau had neglected research and development work in the improvement of consumer appliances and equipment in order to concentrate on filling orders for outside organisations.[114] Since orders from outside organisations were probably a better source of income for the bureau and very likely saved money for the ministry, it was most likely ministerial policy to encourage such orders. Probably this was a case where the PPO secretaries, faced with conflicting demands, made the wrong choice. Similarly, in 1976 the Committee for Party Control criticised the secretary of the party committee of the Kharkov Aviation Institute for taking no action when the rector employed twenty more technical and administrative workers than the establishment allowed.[115] This is something which in some circumstances would go unnoticed or be permitted.

For this uncertainty PPO secretaries tend to blame higher party authorities, specifically their immediately superior *raikom* or *gorkom*. These party organs are supposed to provide formal training for new party secretaries, through seminars and conferences,[116] as well as on-the-spot assistance from their instructors. However press reports indicate that all too often such instructors appear only at election meeting time, primarily to ensure that the 'right' candidates are elected to the party buro or committee. More concrete advice and, perhaps more importantly, support in disputes with management are far from forthcoming.[117]

This lack of support should be put in the context that the authorities have no intention of granting PPOs dominant power over management or guaranteeing them the support they feel they need on all occasions. It is an essential aspect of the PPO system, with its built-in ambiguities and essentially inferior position of the PPO, that any PPO–management dispute be brought to the attention of the higher party authorities, who will then react flexibly in accordance with the demands of the general policy line and specifically local features.

Sometimes the failure of higher party authorities to support a PPO results from a conscious desire to maintain a balance between the party organisation and management. In 1972 a party member from the Central Research and Design-Construction Institute for the Preventive Inspection of Pneumatics wrote to *Pravda* complaining that the institute's director opposed the party organisation's use of the right of control. A party meeting had elected a commission to check the scientific activity of the institute and had found serious shortcomings in the work of a deputy director. The latter was removed from his post. But then the *gorkom* punished four members of the commission with reprimands. In the same issue of *Pravda* a letter was published from the scientific secretary of the institute who complained about the one-sided nature of the commission's work, saying that it had decided to make an example of the deputy director. Although there were admittedly some shortcomings in his work, the commission had concentrated on him at the expense of an examination of anything else in the institute. It had proved necessary to refer the matter to the *gorkom* after the party buro had done nothing about the commission's mistakes. A little later *Pravda* summed up the conflict. It criticized the PPO for its poor direction and supervision of the work of the commission. The commission had indeed concentrated on the deputy director to the exclusion of all else, when the case of the deputy director had already been decided and his dismissal was in hand. *Pravda* generally found in favour of the *gorkom* in its opposition to the commission, although declaring that it had somewhat over-reacted.[118] This case is interesting in that it appears to illustrate a local party organ attempting to maintain a balance between functional and dysfunctional conflict. The main question was one of degree of reaction, rather than dispute over facts, and it was the *gorkom*'s task to establish the correct degree of reaction.[119]

However there would often appear to be more sinister reasons for higher party authorities failing to support a PPO. The Soviet phenomenon of 'familyness' thrives at local levels, and the top managers of local institutions and enterprises are more likely to be members of a 'family' than a part-time PPO secretary. In 1982 *Pravda* described in some detail the case of the Krasnodar Research Institute of Agriculture, in which the director and *raikom* first secretary had protected a dishonest deputy director from the attacks of the PPO. Eventually the party secretary was not reelected to his post, ostensibly because he wanted to write up his dissertation.[120] One feature of the Soviet system which gives great strength to 'families' is its 'culture of authority'. Institutional leaders represent supreme authority in their bailiwicks, and there must be a presumption of their infallibility, as there is a presumption of the infallibility of the highest authority, the party leadership. This is a presumption on which local party leaders rely in protecting their 'family'. One of the best examples concerns an administrative, not research organisation. In 1975 at an open party meeting in the Kirgiz Farm Machinery Association (*Kirgizsel'khoztekhnika*) strong criticism was expressed of the first deputy chairman.[121] The PPO secretary chaired the meeting, while a member of the Kirgiz Central Committee's Agriculture Department was present. After the meeting the Central Committee official and the deputy chairman attacked the party secretary for allowing criticism of a leading official at an open party meeting.[122] Thus, through a combination of system-wide legitimacy and local 'family' reasons PPO secretaries can by no means feel certain of the support of higher party authorities.[123]

Another problem PPO leaders face is their possible lack of expertise in all the areas in which PPOs operate. No one could expect to be a specialist in ideology, scientific research and science management, and particularly not the part-timers who make up the bulk of research

PPO secretaries. The most regular complaints of lack of competence are in the ideological sphere. Research PPOs, particularly in the natural sciences, are unlike *vuzy* in having no pool of professionally trained ideologists. Thus even attempts to improve the quality of PPOs' ideological work by putting the emphasis on deputy secretaries for ideology have foundered on the difficulties of finding suitable personnel. The Oktiabrsky *raikom* in Moscow was so concerned about the problem and the resulting high turnover in the position, that it decided to include the deputy secretary for ideology in twenty major institutes on its *nomenklatura*. It is hard to believe the post was not already there, and the move would hardly solve the problem anyway.[124]

The other main difficulty arises from the enormous stress in recent years on complex schemes of science management, the *naukovedenie* revolution. The attestation, productivity and incentive schemes that have been introduced are bewildering in their complexity. It is claimed, entirely believably, that not even the specialist planning and accounting departments of institutes can cope,[125] while demands for economics training for researchers and managers are insistent. There is no reason to believe that the PPO secretary will be better off in this regard than those he is supposed to be controlling.

Motivation of the PPO secretary

In the discussion so far I have assumed that the PPO and its secretary have wanted to use all available powers. Problems have arisen either because the PPO secretary did not have, or did not feel confident of having, sufficient power, or did not know how to use it, or because he did not receive support from higher authorities.

But perhaps a more serious problem is that PPO secretaries do not in fact want to exercise the powers available to them and would prefer to keep a safe distance from higher party authorities. The ideal party secretary is the 'good party man', who will take action on his own or go to higher party authorities when he feels action needs to be taken against either management or the rank-and-file. However it seems very likely that PPO secretaries of two other types will often exist. These are the 'management man' (in Russian the '*karmanny*' or 'pocket' secretary, i.e. one who is 'in the pocket' of the director) and the 'constituency man', i.e. a secretary who feels he should represent the interests of the rank-and-file, both of the PPO and of the institute in general.[126]

The 'pocket secretary' exists because very often it is not in the secretary's own interests to oppose his director. As we will soon see, research party secretaries appear to be nearly always practising researchers involved in party work on a temporary basis. The director is always an important and powerful figure in the scientific world, and a party secretary with an eye to a scientific career would be reluctant to cross him. Further, the secretary, being a scientific man, might share the director's deviant scientific or ideological interests. Thus in cases of conflict between the party and the director, the PPO secretary might well take the side of the director, or simply cover up for him to prevent the party apparatus taking an interest in his activities.

A party secretary who adheres to professional interests more strongly than the director is likely to become a 'constituency man'. He will support his constituency in its demands for such scientific rights as the freedom of scholarly opinion, unhindered contact with other scientists, and general intellectual honesty. In such cases higher party authorities will almost certainly be on the side of management, whether actively or passively, and particularly in those cases when demands for scientific freedom are transformed into demands for ideological and social freedom for the whole of society.

Every party secretary will feel pressures simultaneously from party, management and constituency to behave according to one of these types. In various different circumstances one might have the advantage and be able to exert decisive pressure – the party during periods of an ideological crackdown or strong involvement in economic affairs, management during the periodic campaigns against excessive party involvement in everyday operational matters, the constituency during the rare periods of ideological liberalisation. But here I am not concerned with these external circumstances so much as with the characteristics of the post of PPO secretary and the type of people found in the position. If we look at the background, training and positions, ambitions and motivations of PPO secretaries, and at how much these factors correspond to the demands and rewards – or lack or rewards – of the job, we should be able to determine which type of PPO secretary is most likely to predominate.

Before examining the data we should look at exactly how a PPO secretary arrives in his position. If one of the main actors involved, the party, management or constituency, has an overwhelming voice in the choice of who will be secretary, his type will be virtually predetermined. Like all executive positions in the party apparatus the post of

PPO secretary is elective, although again like all executive party posts not directly so. The rank-and-file party members vote directly for the party committee or buro, the members of which then elect their secretary from among themselves.[127] The election of the buro or committee takes place at the election-report meeting of the PPO, a general meeting open to all party members in the collective.[128] The size of the buro or committee is set by an open vote of the meeting before the election.[129]

Nominations are then called and accepted until a motion is moved and passed that the list of candidates be closed. (The motion can be moved from the chair.) Discussion is next called for on each candidate in turn. If no criticisms are made of a candidate he is automatically included in the list to be subject to ballot. If criticisms are expressed, an open vote of the meeting decides whether the candidacy should be accepted or rejected.

Voting is secret and voters have the right to cross out, i.e. vote against, any candidate on the list, and add new names if they wish. Anyone receiving the votes of more than 50 per cent of the party members at the meeting is deemed elected. If it happens that the resulting number of elected members is more or less than the number of members originally decided upon, an open vote is taken on whether to accept the new number of members or hold a new election.

Because it is at this stage that the first moves in the election of a popular party secretary must be made and the moves against an unpopular one will usually begin, the *raikom* will be concerned to ensure that things go the way it wants. There will almost always be a *raikom* representative in attendance, who will usually present a list of recommended candidates. The source that has been relied upon in this section states:

> It can happen that the *raikom* or *gorkom*, with the aim of improving the work and strengthening the leadership of a primary party organisation, will consider it necessary to transfer to it an experienced and capable worker from another party organisation. That the *raikom* or *gorkom* has such a right in unarguable. It is the duty of communists to regard the suggestions and arguments of the higher organ in the most responsible way.[130]

The *raikom* will make such recommendations for candidates from within the PPO as well, and the confidence of the party press that such recommendations will usually be accepted is undoubtedly well-founded.[131]

However some tact is necessary and expected. One *raikom* Organizational Department which before meetings gave its representatives a chart including a section listing recommended candidates was criticised by the *obkom* for breaching intra-party democracy. The *raikom* defended itself by saying that the chart was only a guide for its representatives, that it was never displayed at meetings, and that the party membership was under no obligation to vote for the listed candidates. Nevertheless *Partiinaia zhizn'* supported the *obkom*, although adding that it was sometimes necessary for a *raikom* to recommend a candidate.[132] Presumably intra-party democracy demands that the list of recommendations be committed to memory, not written down!

It is not in the interests of the *raikom* to risk any revolt at the election meeting, or to promote conflict in an institute by forcing the election of a candidate unpopular with either the management or the rank-and-file. Sometimes a *raikom* finds that it has misjudged the mood of the collective or finds itself committed to open conflict. It is then likely to find itself in a situation where its own candidate is ballotted out or an unacceptable candidate voted in, in what is after all a secret ballot. It is then forced either to give way or to adopt various illegal measures to overturn the election. The party press often contains criticisms of *raikomy* who have adopted the latter course; neither improve the credibility or legitimacy of the *raikom*.[133]

Thus even in the most difficult situations a wise *raikom* will sound out opinions in advance and try to arrive at some consensus or compromise choice. Take, for example, Nekrich's description of the 1965 and 1966 elections in the Institute of History. In 1965 the *raikom* could not accept the 'liberal' Tarnovsky as secretary, but compromised by suggesting Danilov, perhaps less liberal but certainly no reactionary. In 1966, long before the election was due, the *raikom* several times called in the party secretary, director and other members of the institute staff to discuss potential candidates. The director lobbied for a new secretary, but was defeated, perhaps as much by the refusal of his candidate to stand as anything else. Danilov also managed to convince the *raikom* that to oppose the reelection of Nekrich to the party committee could only noisily fail. When the meeting took place, although members of the outgoing committee were subjected to strong personal attacks, the *raikom* representative made no effort to influence the outcome of the election and the old committee was returned.[134] While the situation was not ideal for anybody, scandal was avoided and the prestige of the party was probably higher than it would have been otherwise.

Thus all the actors involved in the election of a PPO secretary are in a position to exercise influence over the election. The rank-and-file can add any name they like to the ballot and will probably have their feelings tapped beforehand. However out of a desire for a peaceful life and some degree of commitment to the party's will, the party membership will usually respect the wishes of the *raikom*. As for the director, he can lobby strongly for his candidate in the *raikom* and among the various factions that usually exist in Soviet institutes.[135]

What sort of people are elected? I have collected some biographical data on the secretaries of research PPOs. The sample is small and the methodology far from foolproof,[136] but the data are supported to a large extent by descriptive information from Soviet and emigre sources, both published and unpublished.

I have identified 144 PPO secretaries (some deputy secretaries are also included), and have found information on 63. (No statistical significance can necessarily be attached to the number of secretaries on whom no information has been found, since some have been identified only recently, and no search has been carried out under their names.) The information usually includes the number of articles they have published, the academic degree they hold and other posts they have occupied. Thus some idea can be gained of a secretary's scientific standing and productivity (with the warning of course that the dishonest publication of articles and awarding of degrees is endemic in the Soviet Union) and his career interests and prospects.

One question that cannot be answered from the data so assembled is the number of full-time research secretaries. According to the Party Rules a PPO with 150 members or more is entitled to a full-time secretary. We have no data on the number of research PPOs of that size.[137] The survey reported by Miller shows 17 per cent of Academy institute party secretaries as full-time, and 46 per cent of branch secretaries.[138] I might have expected the branch figure to be lower, but not on the basis of any statistical data.

In fact whether the secretaries are full-time or not is not the main question. More important is whether they are professional party workers or researchers. My evidence indicates the latter. The party considers it important that PPO secretaries be close to their constituency, that they know personally the people they are dealing with, and that they have sufficient knowledge to understand the institute's work. The impression gained from Nekrich's biography (and he worked in a field where it would be relatively easy to find an adequately knowledgeable party professional) is that there was a limited number of people in the institute who were suitable for the job and willing to take it on, and that those people more or less took it in turns. The same names appear throughout the 30 years he is describing and they are all professional researchers in the institute. Of the 63 secretaries on whom I have information, 34 regularly publish in scientific journals, 41 have academic degrees in the relevant field, and 21 have been identified as holding research positions before, during or after their term as secretary.

There are two possible consequences of PPO secretaries being drawn from the ranks of professional researchers. Firstly, as described above, one might expect the party secretary to share the professional interests of his research colleagues. This might be a belief in a proscribed branch of science – for example, the condemnation of the party secretary of the Institute of Plant Breeding in 1937 along with his director Vavilov for their pro-genetics activities – or a general desire to promote such scientific values as freedom from administrative control, access to foreign information and researchers, academic recognition on the basis of merit, etc. Here Danilov of the Institute of History in the mid-1960s is a good example. He supported his PPO, particularly its committee, and most of the rank-and-file researchers in their conflict with the institute director.[139]

The second possible, and perhaps more probable consequence, is that any professional scientist interested in a research career would be reluctant to cross his director. Institute directors have enough influence over work references, allocation of research projects and promotions, not to speak of such mundane matters as bonuses, trips abroad and access to rest homes, that no researcher planning on staying in the same institute or even discipline would willingly make an enemy of him. This predisposes those who would not like to be in his pocket to avoid the job, leaving the field clear for those who gladly accept the temptation of overly close relations with the director.

Either way, commitment to professional interests or to the director, the interests of the party will not be supreme. Further, professional researchers have no particular training in ideology (unless of course they are in a social sciences institute). In higher education, even in natural science faculties and *vuzy*, there are always departments of social science providing obligatory instruction in Marxism–Leninism, scientific communism, etc. Thus there is always a pool of professional ideologists, and a large proportion of party secretaries in educational institutions in fact come from this pool. Research institutes have no such pool.[140] Of course it requires no great training to understand what sort of social behaviour is unacceptable in the Soviet Union, there is rarely great subtlety in the application of ideology to science itself, and *raikomy* run regular seminars for PPO secretaries to bring their ideological knowledge up to date. But it is not a lack of ideological knowledge or expertise that is the problem, so much as a lack of people who by their very choice of profession have shown themselves already prepared to accept all the implications of Soviet ideological control. Someone who has chosen to train and work as a teacher of Marxism–Leninism is more likely to be prepared to accept the intellectual compromises and behavioural demands of the state ideology than someone who has chosen to train and work as a theoretical physicist, biologist or even historian. The implications for the relative degree of commitment to the party line are obvious.

The next aspect of PPO secretaries to be examined is their prestige and status as scientists. The party puts great stress on the need to have highly educated secretaries and party buro or committee members. Great publicity is given to the increasing number of secretaries in all types of PPOs with higher education. In research institutes the concern is to have a good proportion of candidates and doctors of science and researchers of generally high status and prestige. These people will not only be better equipped to understand and evaluate the work being done in the institute, but will also be taken notice of when they express an opinion on scientific or even social matters. Kozlova describes the improvement in academic status of PPO secretaries in Moscow research institutes. In 1971, of 20 party secretaries in Academy institutes in Cheremushki *raion*, 3 were doctors of science and 15 were candidates; by 1977 7 were doctors and 13 candidates. In 1974 in the *raion's* research institutes as a whole, 27 party secretaries out of 39 had advanced degrees. In 1978 in the Moscow institutes of the Academy, VASKhNIL and GKNT, about 20 percent of secretaries had doctorates and over 70 per cent candidate degrees.[141]

Of the 42 secretaries I have identified with academic degrees, 31 have candidate degrees, while nine are doctors of science. The list includes one winner of a State Prize for physics and a corresponding member of the Kazakh Academy of Sciences. With 34 of the secretaries also involved in the regular publication of articles, a good proportion of research scientists are active scientists. The major Academy institutes seem particularly likely to have apparently prestigious scientists as PPO secretaries. For example, A. A. Borisov, secretary of the PPO of the Institute of Chemical Physics of the Academy of Sciences in 1971, is a doctor of chemical sciences and a regular author of scientific articles; V. I. Lakomsky, a doctor of technical sciences and also a prolific writer, was the secretary of the PPO of the Paton Institute of Electrical Welding in 1969; L. M. Liamshev, party secretary in the Institute of Acoustics in 1971, is a doctor of physico-mathematical sciences and very well-published. A. A. Sergeev is a well-published doctor of chemical sciences who was party secretary in the Institute of Elementary Organic Compounds in 1972, while V. P. Silin, secretary of the PPO of the Lebedev Physics Institute in 1971, is the author of many articles, a doctor of physico-mathematical sciences and a winner of the State Prize. The noted academician N. M. Emanuel' was, according to his obituary, often elected party secretary of the Academy's Institute of Chemical Physics.[142]

Of course the party wants a prestigious scientist to be listened to by his colleagues only when he is saying what the party leadership wants him to say. But a highly prestigious scientist is perhaps more likely to feel able to resist party pressure and express those professional interests which should be particularly close to his heart than someone with a lesser reputation. Nekrich clearly attaches some importance to the fact that the 'liberal' party committee elected in the Institute of History in 1965 contained highly qualified professionals, including eight doctors of science.[143] In a personal communication he has informed me that when the party committee was replaced in 1967 the majority of new members were candidates of science and junior research workers, his implication being that the less qualified the party committee and secretary, the less likely they are to be 'liberal' and independent. This includes, of course, independence from management as well.

An important factor here is that a researcher with an established reputation will be less seriously affected by the demands on his time of being PPO secretary. It was said above that most research party secretaries are not full-time officials, and on a part-time basis it can be a most time-consuming job. Presumably allowances are made when it comes to formally measuring work performance for determining bonuses and ratings at reattestation, but this is not enough for a young researcher struggling to make a reputation for himself at the forefront of his field. There are no figures on the amount of time a research PPO secretary is likely to spend on party work, but all my informants assure me that it is a heavy load. Nekrich comments specifically on the advantage non-party researchers have over party members in this regard.[144]

This is confirmed by the publication records of party secretaries. A number of the secretaries I have identified who are generally prolific writers of articles have ceased publishing during their term as secretary. Thus D. B. Dunaevsky, described as party secretary in the All-Union Research Institute of Leguminous Plants in 1970, had articles published in 1965, 1966 and 1967, none in 1968, one in 1969 and none in 1970. One could assume his term ran from 1968 to 1970, a normal period. V. A. Sergeev was spoken of as the secretary of the party buro of the Institute of Elementary Organic Compounds in 1972. His previously regular articles stopped appearing in 1969. P. I. Iushkov has never been an overly prolific publisher of articles, but they have appeared regularly under his name except for the years 1967 to 1969. He was reported to be party secretary in the Institute of Plant and Animal Ecology in 1968. In 1963 B. A. Zhubanov was identified as deputy secretary of the party buro of the Institute of Chemical Sciences of the Kazakh Academy of Sciences. He published articles in 1962 and 1963, but none in 1964 or 1965. From 1966 he became a very prolific writer.

An established researcher might be more prepared to put up with these shortcomings of the job than a younger scientist seriously interested in a scientific career, and there is indeed evidence of relatively eminent scientists taking on the job. But perhaps the more likely reaction is that it will be avoided by any serious scientists, established or not.

There is some indirect published evidence that the position is not liked. The handbook cited above on party elections states slightly ominously: 'Active social participation is the duty and obligation of a communist. Therefore the personal agreement of a party member to his election to a leading organ is not necessary.'[145] The party member does have the right to declare his unwillingness to be a candidate, but it is then up to the party meeting to decide by an open vote whether to accept the withdrawal or not.[146] The fact that such a rule exists is in itself an indication that people will avoid the position if they can. Periodically specific cases appear in the Soviet press. Although none of them concern research PPOs, it is worth mentioning one of them as an example. In 1964 *Partiinaia zhizn'* received a letter from the secretary of the PPO of the Cheliabinsk regional office of Gosbank in which he rather hopefully suggested that the party members had made a mistake in reelecting him since he was unable to control the behaviour of his son and was therefore an unfit person to hold such a responsible post. The journal supported the reelection, since in its opinion the secretary was doing his best with his son. It was stated drily that he should not try to avoid his social obligations.[147]

All the indications are that the secretaryship is a position to be avoided by anyone who wishes to concentrate on scientific work and who does not wish to find himself in a potentially embarrassing or dangerous relationship with his director or colleagues. The formal obligation to accept the position might mean some 'good' candidates are put forward and elected. A good scientist, committed to the Soviet system but with some degree of conscience and therefore credibility with his colleagues – the type of person favoured by the party – might not want the post but must accept it if it is offered. Such people of my own acquaintance also have a sense of duty which causes them to accept the job with more or less good grace.

Nevertheless one cannot help feeling that in a situation where true scientists seek to avoid the position, the ground is clear for more self-serving candidates. Official Soviet sources naturally do not touch on the issue, but it seems probable that many PPO secretaries are scientific non-entities who are hoping to help their career along by ingratiating themselves with the powers that be, specifically the director.[148] Emigre writers state that it is common for a party secretary, once he has finished his term, to be given a promotion by the director as a reward for good behaviour. Agursky implies that the promotions given retiring party secretaries in the Experimental Research Institute for Metal-Cutting Lathes were not a reward for scientific ability. He cites one particular case of a secretary who on completing his term became deputy director 'for the development of the branch', a post with no clear functions.[149] Such promotions are important, since the increases in salary are dramatic as one progresses from junior research worker to deputy director.

However there is little other evidence that a term as PPO secretary seriously helps one's career, whether that career be scientific, administrative or party. While I have not had much success tracing the subsequent careers of research party secretaries, having identified a subsequent post in only seven cases, the evidence is not strong that PPO secretaries go on to noteworthy careers. The seven include a deputy director,[150] chief agronomist of an agricultural production administration (not a research post), a department head (twice), and a sector head. In the twenty cases where I have identified posts held before or during the term as secretary the posts are nearly all at sector/laboratory/department head level. Thus posts held during and after terms as secretary do not differ significantly. Two important exceptions are V. V. Shiriaev, who went from being party secretary in a machine-building institute to first secretary of Moscow's Proletarsky *raikom*, and Viacheslav Kudinov, who moved from party secretary of the giant Paton

Institute of Electrical Welding to a deputy chairmanship of GKNT.

It is particularly interesting to compare the posts gained by research party secretaries with the important posts subsequently occupied by so many of the party secretaries of *vuzy* I have identified.[151] Although they have been subjected to the same search procedure, the results for the two categories vary considerably, with the subsequent careers of *vuz* party secretaries being far more impressive.

It appears probable that a term as party secretary in a research institute does increase one's chances of gaining a reasonably secure and well-paid post within the confines of one's own institute. But this is not going to be enough for a truly ambitious person who wants to go beyond the deputy directorship of a research institute. I have found little evidence that a term as PPO secretary helps one get beyond this point. In fact the reverse is perhaps more likely – one is kept from one's research for a number of years and is put in a position where conflicts with various people important for one's career are not unlikely. The result is that, with the rare exception of those periods of relative ideological freedom when committed liberals might become interested, the only people who are genuinely attracted to the job are ungifted researchers wanting a secure existence but without any great ambition, or old and failed party hacks who have retired to the peaceful life of research. In either case the main desire would be for a quiet and unthreatened life. Certainly there would be no desire to represent the interests of one's professional colleagues, nor would there be any desire to engage in great struggles with the director over, say, ensuring complete honesty in planning and accounting procedures. Thus one should not expect any strong commitment to those aspects of PPO control of research that have received considerable attention from party authorities in recent years. The conditions would seem to be perfect for the breeding of 'pocket' secretaries.

The nature of the job – time-consuming and of limited career value – and of the people found in it – professional researchers, usually middle rank, of limited ambition and perhaps talent – lead one to expect that the 'management man' type of PPO secretary will predominate in normal circumstances. The party might try to offset that by finding as a candidate a more prestigious scientist, sufficiently widely respected and perhaps honest to be able and prepared to resist management when necessary. The danger then is that such a secretary will come to represent his constituency too zealously, particularly at times of ideological ferment. Thus the party finds itself caught between two evils – a secretary beholden to the director, in my opinion by far the more common case, or a 'liberal' secretary representing the rank-and-file. Does the party have any way out of this impasse? Is there any way it can reliably find candidates that will represent its interests in Soviet research institutes? One might expect as a response the party professionalisation of research PPO secretaries, making them committed party workers with a career in the party apparatus, as has happened to some extent in the PPOs of ministries and major state agencies and even those of *vuzy*.[152] But as yet there is no sign of this happening in research institutes. I doubt that it will. The party might sometimes have difficulty finding the ideal secretary, and this must interfere with the important role that the party wants PPOs to play. But to professionalise the post would needlessly antagonise the director. It would also replace the present problem of lack of expertise of party secretaries in ideology and science management with the problem of lack of knowledge of the scientific process itself. It would also remove that close contact with the collective which is such an important part of party control in research institutes. After all, most 'liberal' researchers are very good at reading the signs and removing their 'liberal' principles from public view as the ideological atmosphere changes, even without any great pressure from the authorities. That is certainly the lesson to be gained from Nekrich's Institute of History.

The evidence suggests that the PPO has considerable powers of control over both management and the rank-and-file workforce, these powers being embodied primarily in its right to vet all personnal appointments, to apply party disciplinary measures to that large proportion of research personnel who are party members, and to exercise the formal provisions of the right of control. One might expect these powers to be adequate to maintain ideological and political order among the scientific intelligentsia, to prevent institute management engaging in illegal behaviour, whether based on corruption or incompetence, and to ensure the quality and success of research and that the party's science policies are implemented. In fact many specific cases and the recurring problems that can be identified in Soviet R&D suggest that PPOs regularly fail in their control functions. This is perhaps not surprising when one considers that they are deliberately given inferior status and power to institute management, that their party superiors are often unwilling, whether through deliberate policy or indifference, to provide essential support, and that PPO leaders are often strongly motivated not to exercise their powers, at least in the way that the party authorities might wish. PPOs are undoubtedly a useful tool in the party's management of science. But they are perhaps a disturbingly unreliable 'ultimate device of control and management'.

Conclusion

The conclusion contains two main sections. Firstly, there is a summary of my empirical research on the involvement of the CPSU in the management of Soviet science; secondly, there are some comments on the implications of the findings of the empirical research for the continuing debate on the best theoretical approach to the study of the Soviet political system.

EMPIRICAL FINDINGS

The previous chapters have each been devoted to a particular level of the party apparatus – the PPO, local and regional party organs, and the Central Committee and Politburo. In this Conclusion I will approach the role of the party from a functional perspective. There are four main aspects of party control of Soviet science, as indeed of all elements of Soviet society. Firstly, there is the maintenance of ideological control, which can be taken to include control of the political behaviour of scientists in their private and social lives as well as their scientific work; secondly, control over personnel appointments; thirdly, the determination of major policy directions; fourthly, control of the implementation of policies and decisions. These four functions are what the party itself sees as the components of its 'leading role', and they appear to be an excellent framework for our own evaluation.

Ideological control

Marxism–Leninism is claimed to be a universal and overarching science, able to provide a framework for all other natural and social sciences. At various times in the past that framework has been a tight and constricting one, with devastating consequences for many branches of Soviet science. In the social sciences the baleful consequences of the ideology continue to be felt, but it could be argued to a far lesser degree than under Stalin. This is well illustrated by such ideologically sensitive social sciences as economics and sociology. The intricacies of the politics of these two disciplines are, I fear, beyond human understanding. But we clearly see, on the one hand, their continuing if variable subjection to ideological strictures and their extreme vulnerability to the attacks of ideological conservatives, but, on the other hand, the enormous expansion in their permitted fields of study and their resilience in the face of attack. Thus mathematical economics and sociology had their greatest victories over the ideological conservatives in the late 1960s, at the time of a severe crackdown on political dissidence in the scientific world. The conservatives gained their victory in the early 1970s, with the downfall of Rumiantsev and the purges of *IKSI* and the non-political economists in the Institute of Economics. However the mathematical economists' institute, *TsEMI*, retained its influence, while in 1975 the chief purger of *IKSI*, Rutkevich, was himself purged. Both mathematical economics and sociology live on, somewhat dangerously, as reasonably active and influential disciplines.

The natural sciences are in a better situation. One gets the impression that the party authorities learnt their lesson from Lysenkoism and are determined that such a thing will not happen again. It is well understood that the costs to Soviet science would be too great. If the scientists are confident that ideological and political power will not again be used in a Lysenkoist way, it is indeed an important change in party–science relations, one that is likely to increase the feeling of independence of the scientific community, but at the same time perhaps improve its relations with the party.

In many cases the tendency to an improvement in relations is nullified by the party's total determination to control the political lives of all Soviet citizens, including scientists. The scientific community, with its more than proportionate contribution to the ranks of the dissidents, has borne the brunt of the crackdown on ideological non-conformity since the second half of the 1960s. This has presumably had a negative, if unquantifiable, effect on the morale and motivation of scientists.

But of perhaps greater significance as far as scientific morale and morality are concerned is the whole 'culture' of political control that exists in Soviet society. When combined with a related obsession with bureaucratic regulation, this 'culture' leaves far too much room for intrigue, the settling of scores and the furthering of ambitions by non-scientific methods. Although scientists who are tempted to engage in such activities now have to do so largely without recourse to ideology or to the ultimate weapon of top-level political support, the opportunities for denunciation, arbitrary dismissal or removal of bonuses and fringe benefits are enormous. While these measures are not now applied against whole scientific disciplines, they are applied to scientists as they do their work. The oppressive atmosphere that results can only do great harm to the morale and morality of Soviet scientists.

Personnel

However, these powers can usually be used only by those in positions of managerial authority, and therefore we have to look closely at the power of the party to put 'its' people in managerial positions. There is no doubt that the party has total power over personnel appointments. Although *nomenklatury* are never published in sufficient detail that we can be sure of the levels of the party apparatus at which various appointments are confirmed, some intelligent guesses can be made. PPOs have the power to vet any personnel movements within an institute, while managerial positions (laboratory and departments heads, directors and their deputies in minor institutes) are checked by middle-level party organs, or the Central Committee apparatus in the case of particularly sensitive institutes. Directorships of major institutes probably go to the Central Committee apparatus, while the top positions in the major management bodies – the Academy and GKNT – are probably within the provenance of the Politburo.

This personnel power leads many to conclude that the party appoints to managerial positions those who share its values of tight dictatorial control, or that it can at least force its appointees to act according to those values. While modern-day examples of unscrupulous scientists using administrative measures for their own ends appear to be exaggerated (for example, plate tectonics), it is nevertheless probable that the creative initiative and energy of Soviet scientists are dampened by the stifling atmosphere of political and bureaucratic repression. In the opinion of this author, more work needs to be done on the role of institute directors in Soviet science. However it appears that in many cases they are 'party' people whose rather unsavoury reputations are well-earned.

Nevertheless this line of reasoning, if not the conclusion, has to be modified to some and perhaps an important extent. The authorities did not reject Lysenko out of a passionate attachment to the purity of science – they did so because they could not fail to recognise the practical damage that Lysenko had brought to Soviet science. Similarly, some allowance has to be made today by the authorities for the fact that scientific bullies can do enormous damage to performance. Throughout Soviet history there have been scientists in positions of authority who clearly have not shared the values of the party, and still the party has no choice but to admit that at times scientific quality has to come before a total commitment to a 'party style of work'. In selecting science managers, whether they be the heads of the major management bodies, institute directors or PPO secretaries, the party faces the dilemma that it has to find people who are entirely reliable in its terms, but who have some scientific talent (in today's specialised scientific world that might not be easy), and, perhaps most difficult, who can maintain at least a minimum of respect among their colleagues. The reality of the dilemma can be seen in the ambiguous behaviour and reputations of so many Soviet science managers.

Overall one does not get from *samizdat* and emigre literature the impression that Soviet science is run entirely by 'party hacks'. Many of those running it might be of dubious moral character, but they are usually given credit for cunning, organisational ability, diplomacy when needed, and a realisation that their careers depend on running a successful scientific outfit. That is something which can be done only by meeting their subordinates at least half way. While they probably could be relied on by the party in a crisis, they, and indeed the party, would rather that crisis never arose.

This is not to deny that the quality of Soviet science most certainly suffers from overbureaucratisation, excessive political control, particularly of international communications and scholarly exchanges, and from poor morale among scientists. How important these problems are next to others such as infrastructural shortcomings and the production priorities of the regime is very difficult to say.

Policy-making

There is no denying that the Politburo is the major policy-making body in the Soviet Union, and we know that important decisions on scientific and technological matters are made at that level. Further, there is good evidence that all top party leaders have shown a strong awareness of the importance of scientific and technical matters. In the present Politburo there are a number of members who appear to take a special interest in science and technology.

Despite all this there is no reason to believe that the Politburo has a major initiating role in science policy, whether we are talking about setting scientific priorities or changes in science management procedures. These are not seen as matters of absolutely first-rank political priority, and often require such technical expertise that the room for Politburo involvement must be limited.

There is evidence, although it is not absolutely contemporary, that top scientists have direct access to major political figures, and that in some cases they are able to exercise considerable influence over decisions in their fields. The best example is Igor' Kurchatov, who as a result of his success in heading the Soviet nuclear bomb programme, built up considerable political power, with particularly close links with Khrushchev. This meant not only that he had influence over Khrushchev in matters on the boundary between science and politics (the declassification of fusion research, the sponsoring of radiation genetics research and attempted influence on nuclear testing policies), but that he had enormous power to determine research and funding priorities in Soviet physics.

One imagines that more routine decisions, and most of the politics of Soviet science policy-making, are centred in the state apparatus, presumably within the Council of Ministers. The heads of Gosplan and GKNT, two major bodies with responsibilities for science, are both deputy chairman of the Council of Ministers and therefore have seats on the presidium of that body. In what is often an antagonistic relationship Gosplan would seem to have the greater political clout. It is probably for this reason that GKNT has closer relations with the Academy of Sciences than one might expect. The Academy, despite having statutory power over the coordination of all fundamental and social science research (and at least consultative power over the coordination of all research), does not have formal representation on the Council of Ministers or its presidium. This could well be an incentive for the Academy to maintain good relations with GKNT. It is likely that GKNT prepares many of its submissions to the presidium of the Council of Ministers together with the Academy. The Academy's lack of formal status in the Council of Ministers has probably also spurred it on to develop its contacts at the highest levels of the party apparatus (probably routinely at Central Committee secretariat level, the responsible Central Committee secretary being Zimianin. One suspects that the president of the Academy has reasonably ready access at Politburo level.) This would be a contact designed to ensure that the Academy's views are known at the highest levels.

Given this, the lack of evidence of the involvement of the Central Committee apparatus in science policy-making is somewhat surprising. But with the Science Department essentially an ideological body and the branch departments perhaps dominated by production concerns, there would seem to be little room for such involvement. If such a finding is correct, the direct relationships between scientists and top political leaders become even more important.

I am impressed by the admittedly circumstantial evidence that the party is keen to encourage the Academy in its activities, and that indeed the party sees the Academy as a weapon to be used against the technologically indolent branch ministries. The role of the Academy of Sciences in R&D has been steadily increasing since the blow Khrushchev dealt it in 1961; Academy managers speak endlessly of the support they get from the party in the exercise of their power (not something which is traditional in Academy publications); and it is made quite clear that the branch ministries, and by implication Gosplan, are the main obstacles to technological advance. This ever-increasing role of the Academy and the shifting of its responsibilities ever closer to the production end of the spectrum are signs of the degree to which the party feels it depends on the Academy for the technological progress which it sees as essential. With great risk of exaggeration one could say that there is an alliance between the party and the Academy, with GKNT perhaps its spokesman in state bodies, against Gosplan and the ministries. The potential power this alliance gives leading scientists is operationalised in the major role they play in drawing up research programmes (the *de facto* power that comes from 'planning from below' plus their major role in drawing up and managing long-term R&D programmes).

There is little evidence to suggest that the party has a monopoly of Soviet science policy-making or that it regularly imposes its policy will on the scientists. Indeed the scientists appear to have major opportunities to involve themselves in the science policy–making process. Further, there are signs of their cooperation with the party, not entirely from a position of weakness, in this.

Implementation of policy

It is in the implementation of policy that one would expect the party apparatus to come into its own, with a territorial network spread throughout the country and with primary organisations in every research organisation. The Central Committee apparatus is said by many virtually to run the country, and regional party secretaries are considered to have almost dictatorial powers in their regions. PPOs also have considerable formal powers over management. This should enable the party apparatus to guarantee the implementation of science policies, both ensuring that the scientists take on the tasks that have been set and providing 'expediting' assistance to them in fulfilling those tasks.

The Soviet Union is such a bureaucratic society that it is not surprising that tasks can be fulfilled in purely formal terms relatively easily. Thus claims that the party apparatus successfully controls the fulfilment of research plans can be taken initially at no more than face value. It is interesting that in those very high priority projects where fulfilment must be in more than formal terms the involvement of the party apparatus seems to be minimal (for example, the bomb and space programmes. Even in more routines defence research the party presence seems to be less than in civilian research).[1] In more routine research, where verification of plan fulfilment is claimed to be one of the party apparatus' major tasks, complaints continue of the difficulties in cutting minor research topics out of programmes, getting whole institutes to work in the fields for which they were set up, and persuading or motivating scientists to take an interest in the production application of their work. The party apparatus is constantly enjoined to remove these shortcomings, but one must have doubts about the motivations of the PPOs, the capacity or interest of the regional party bodies (where the bodies set up to verify fulfilment are usually scientists working in their spare time), and the capacity of the Central Committee apparatus. While the Soviet control network is powerful enough in high priority matters, the party apparatus would seem to be to a considerable degree powerless to prevent the routine avoidance of plan fulfilment in anything but formal terms.

The record of the party apparatus in its expediting role is also not a good one. We have very little information on the work of the Central Committee apparatus. This might be simply Soviet secrecy, but it is worth considering that it is in fact a true reflection of reality. As already mentioned, the Science Department, with formal responsibility for Academy and much *vuz* research, is essentially an ideological body, with no strong evidence of expertise or interest in the management of research. The industrial branch departments appear to be dominated by 'production' interests that militate against a strong commitment to technological innovation. None of these bodies would appear to have a strong interest in facilitating the coordination of often conflicting R&D bodies or pressuring production organisations into providing research and *vnedrenie* facilities.

The same general remarks could be made about regional party organs. A large number of the cases we have of regional party authorities playing an expediting role describe their lack of success. This is surprising, and enough to make one wonder whether regional party bodies do in fact have much control over local resources. It appears likely that when local production facilities belong to major central ministries, the local party authorities are by no means guaranteed access to these facilities. One should not exaggerate this – the regional party organs still have a lot to offer the scientific community if they are that way inclined, and particularly if there is a powerful party leader in the region willing to push scientific and technological development (Romanov in Leningrad and Shcherbitsky and some of his *obkom* secretaries in the Ukraine). But the possibilities are by no means as great as one might have expected, and efforts to increase the power of the regional party apparatus by regionalising science management are not yet entirely convincing.

The most difficult question to answer when looking at the party's management of science, and perhaps therefore the most important, is the effect on performance of the general atmosphere of political and bureaucratic repression. Few would deny that the hold of ideology over Soviet science, both natural and social, has been considerably relaxed since the death of Stalin. Many would deny, however, that this represents in any way a sufficient degree of political relaxation within Soviet research institutes to enable us to speak of a science free of the baleful effects of political interference. My own opinion is that the matter depends to a crucial degree on the actual individual scientists in charge of the laboratories, institutes and research management organisations. They set the tone within the research unit and usually it is they who directly bring into it, or resist the introduction of political and bureaucratic excesses. There are many, including those who could not be described as anything but party loyalists, who find it desirable and possible to protect their units from the worst of such excesses. The need for quality research and to some extent scientific tradition provide a counterweight of some force to the Soviet Union's oppressive political and bureaucratic culture.

In terms of what Western writers usually mean by 'science management' a great deal of guesswork must be employed, but reasonably concrete conclusions can be arrived at. Firstly, one cannot possibly deny the ultimate decision-making power of the party leadership.

However, probably relatively few scientific decisions go to the highest levels, while scientists seem to have considerable opportunity to influence crucially those that do. Those that do not would appear often to be the focus of considerable bureaucratic struggle, both within the scientific community and between it and other major economic organisations. The power of the scientific community in these struggles is considerable, but by no means overwhelming. The scientists therefore are fortunate that the party leadership is apparently on their side in their struggles with the more production oriented elements of society, even if that support often seems to go no further than moral support.

One of the difficulties the party faces in providing such support is the relative weakness of the party apparatus, apparently at all levels. The enormous emphasis since the mid-1950s on the role of the party apparatus in facilitating scientific and technical progress has perhaps provided the party apparatus with the necessary motivation, and yet it still clearly lacks the expertise and even, seemingly, the power to overcome the notorious sectional interests of the major ministries. This means that it is particularly in the field of implementation that the Soviet Union's science and technology policies come most to grief.

THEORETICAL IMPLICATIONS

What are the theoretical implications of these findings? How do they fit the various models that have been used to characterise and analyse the Soviet political system?

In recent years in the West there has been an increasing impatience, particularly among Soviet emigres, with what they see as the failure of Sovietologists to come to terms with the repressive nature of the Soviet system. This has produced a revival of interest in totalitarian concepts. This has not meant, however, a return to the detailed 'syndrome' models of the 1950s and early 1960s, in which a very specific list of features was sought and found in Soviet society to justify a totalitarian conclusion. The decline in terror and tight ideological control, and the increased activity of specialist groups, have become too obvious for validity to be restored to those models. The essential features of neo-totalitarianism are described in vaguer terms, but can be condensed to two: the non-political sphere of social life is unable to resist the interference of the political sphere; and that interference will always produce negative results. The political sphere is again described vaguely (and often tautologically as that part of Soviet life which is bad and repressive), but will usually include at least the structures of the party leadership and apparatus. Many who write in this vein however see the division, in terms of individuals, as crossing functional and institutional boundaries. Thus 'partocrats' can be found in the scientific community and 'technocrats' in the Politburo. To this extent neo-totalitarianism, while not necessarily denying the existence of functionally specialised groups, sees them as having no independent status and as being irrelevant to analysis of the Soviet system. The essence of the model is that, at the moment at least, the 'partocrats' and the values they represent dominate the political and, indeed, the entire social process.

In applying the totalitarian model to science two aspects need to be examined. Firstly, does Marxist–Leninist ideology continue to be applied to science in such a rigid and narrow way that political barriers are routinely put around major areas of scientific research; secondly, are 'partocratic' values of political and bureaucratic repression so pervasive and dominant, not just in top-level policy-making but also in lower-level social interactions, that there exists no possibility of science being sufficiently independent to be effective?

In answering the first question an important distinction has to be made between the social sciences and the natural sciences. While the permissible range of research and differences of opinion in the social sciences has broadened enormously since the 1950s, ideological constraints of considerable rigidity and narrowness are still applied in all social science disciplines. There would be few Soviet social scientists who have not personally come up against these barriers. Thus the two conditions of neo-totalitarianism are largely satisfied – the social sciences cannot escape political interference, and that interference, by definition and empirically, must be harmful to the well-being of scientific endeavour.

In the natural sciences the situation is different. There the party authorities have made a conscious effort to remove science from the influence of the more rigid aspects of Marxism–Leninism. Those who today argue the totalitarian point of view tend therefore to look to the second question put above. They claim that the scientific hierarchy is dominated by 'partocrats', people who whether they are scientists or not hold the values of centralised dictatorial control which serve the party's interests so well, but which are so inappropriate to scientific research. Thus the natural sciences, while perhaps free of the worst excesses of Marxism–Leninism, are denied creativity and originality through the oppressive use of dictatorial bureaucratic power by institute directors and other science managers.

I consider that it is not entirely accurate to describe the leadership of the scientific community as all or dominantly 'partocrats' and to lump them together with all other conservative or reactionary people in Soviet society. Leading Soviet scientists are not revolutionaries or dissidents (although quite a number have come close to the latter at times), but to a large extent they do represent what they see as the interests of the scientific community, even if often for selfish reasons. This requires even the most politically conservative of them to compromise with 'scientific' values.

Thus the totalitarian model, while clearly drawing attention to a feature of Soviet science which is of enormous importance, does not leave enough room for a consideration of the maintenance, for whatever reasons, of 'scientific' values and interests, nor of the willingness of the leaders of the scientific community to defend those values.

The next model, or group of models, is the 'vanguard party' model. Here are included such models as Meyer's 'USSR Incorporated', T. H. Rigby's 'mono-organizational society', the latest corporatist models, and, as my generic name for the group suggests, the offical Soviet theory of the 'leading role of the party'. What these models have in common is that they admit the existence of specialised functional groups, with these groups having interests which in some sense are common to all members of each group. These interests determine the behaviour of the groups and their members in the political and social process. However they are not seen as having independent autonomous existence. They are dominated by an overarching ideology and by a 'leading' (*rukovodiashchaia*) party. They are directed, manipulated and 'encouraged' to cooperate in ensuring that in working for their own interests they also work for the benefit of the social interest, as defined by the dominant party.

I wonder whether this description adequately covers some important, and perhaps in the future increasingly important, features of the Soviet system? Essentially, is the party as dominant as the model makes out? While the reassertion of scientific values at the expense of strictly interpreted Marxism-Leninism might be a special case, does ideology even in other areas have any more than an admittedly important blocking role, that is, is it any more a guide to action or a unifying factor in Soviet society? Secondly, does the party leadership show sufficient drive and determination to suggest that it is the leading or dominant force in society (and can its apparent timidity be attributed simply to the senility, incompetence or fear of change of its members)? Thirdly, and perhaps most strikingly, does the party apparatus have the power to enforce the policies that are determined by the leadership?

To be honest, science management is not the best case to use in attempting to answer these questions. The party has never realistically claimed as dominant or direct a role in science as it has in other fields, for example, economic and agricultural management. However that relative lack of a direct role in such an important field as science needs to be explained, as do the seemingly enormous difficulties the party has in arriving at and implementing a useful programme for technological development. Indeed these difficulties in the science and technology field perhaps highlight the difficulty the party faces in all aspects of political, economic and social control in a world increasingly dominated by technology. As Gustafson has suggested, the indications are that the party leadership has realised that its traditional methods of management and control are inappropriate, and yet it has not been able to find an alternative. This produces a reliance on traditional methods, but only in a half-hearted way.[2] Thus the atmosphere of drift that is evident in the Soviet Union today. There is little about current science and technology policy to suggest that the party is able to assert its dominance, and its latest efforts to increase its role, primarily through regionalisation, tend if anything only to reinforce the impression of weakness.

That brings us to the much-maligned 'puralist' model. It differs from the 'vanguard party' model by granting the various groups in society some degree of independence and autonomy. I am not impressed by claims that the party is no more than the forum for competition between these various groups, the resolution of that competition being determined by a mechanistic accounting of relative strengths, or by claims that the party is simply a referee in such competition. The party has interests of its own, and while perhaps not totally dominant is certainly more than *primus inter pares*. It is particularly assertive when issues of political control of the population are involved. Nevertheless it often appears to find it difficult to assert its will. At times in the policy-making area it seems a matter of having no will to assert, while in policy implementation it finds it extremely difficult to overcome obstruction and opposition. Because of these problems, which are perhaps becoming serious enough to acquire systemic importance, the party needs to involve itself in such 'pluralistic' behaviour as coalition and alliance building. I believe that there is good evidence of the party involving itself in such behaviour in the science and technology policy area.

There are some problems with the pluralist model when applied to the Soviet Union. Firstly, pluralism has been applied primarily to the input side of the political process, as a reaction against the output fixation of the original totalitarian theorists.[3] This has yielded interesting but not always convincing results. Accounts of input into the policy-making process from specialist and functional groups reflect a most important aspect of modern Soviet society. However many consider that this input can be treated on such a 'take it or leave it' basis by the party leadership that the groups providing it cannot be considered truly 'pluralist'.[4] To some extent I share such misgivings, although I also agree with Peter Solomon that the fact that the party chooses to ignore or reject a particular piece of policy advice does not deny the importance of the phenomenon or exclude the possibility of significant although not deciding influence.[5] I also feel that much of the input coming from scientists, particularly those who are not social scientists, has to be treated seriously by the party leadership. The expertise of these scientists does give them some degree of 'technocratic' power. This power does not exist because scientists or specialists are able to force the political leadership to accept their point of view. Such power would exist only if the scientists could threaten the leadership with removal from power – something which some groups in Soviet society could perhaps do, but scientists certainly cannot – or with withdrawal of essential services – also hard to imagine in Soviet conditions.

The power rests rather on the leadership's realisation that their own objectives – remaining at the head of a militarily and economically powerful and politically stable state – make it essential that notice be taken of the views of those with specialised knowledge. In my opinion some Western analysts put too much emphasis on the potential ability of the leadership to withdraw this degree of influence from specialists at a moment's notice. The purge of *IKSI* is often quoted as a prime example of how suddenly a specialist group's influence can disappear.[6] Without wanting to dismiss that purge as irrelevant, I am nevertheless more impressed by how well *IKSI* survived it. One does not doubt that if the Politburo ordered the most extreme measures against any group of specialists tomorrow, its order would be carried out. But the crucial question is how likely such a move is, and how likely does it have to be before we must allow for it in our model.

I can see no reason to believe that the authorities will take what in the present Soviet Union would be the extraordinarily radical step of removing the influence of scientists and specialists as a whole or of any one of the specialist groups that provide policy advice at the moment. It is the reason for the improbability of such a step – the determination of the leadership to continue the modernisation of the economy – that provides those groups with their power and influence. It needs to be mentioned that this power and influence extends only as far as the particular specialised knowledge is of relevance. Even the most influential Soviet scientists have been allowed to push their opinions only on issues closely related to their own fields. However I do not see this feature of the Soviet system as being incompatible with a broad understanding of the term pluralism.

It should also be mentioned that the power and influence of those with specialised knowledge need not only be reactive. Many of the documented case studies of group influence in the Soviet Union deal with cases of group reaction to new leadership policies.[7] Scientists appear to have a capacity to initiate new policies – in cases of new fields of scientific advance through a virtual monopoly of knowledge, and in cases of technological development, particularly if defence is involved, through the inexorability of the logic of technological competition.

Some comment is also required on the dominant 'institutional' variant of the pluralist model. By putting the stress on the bureaucratic nature of Soviet pluralist behaviour 'institutional pluralism' is a most useful concept. It is also one which can be quite readily applied to Soviet science. The Soviet scientific community is clearly institutionalised in terms of its members possessing specific and easily recognisable skills, and of its clearly identifiable institutional bases. The institutional weakness of the branch R&D network rather proves the point, in that the far smaller and less generously endowed but highly institutionalised Academy of Sciences dominates science politics.

However two points need to be made. Firstly, an institutional approach by no means requires that a particular group in society be monolithic. The scientific community might have very general common or universal interests, but it is by no means monolithic. It has a number of institutional bases – the Academy, GKNT, the R&D sections of the ministries – which have varying interests. But even within such institutions there is a great range of opinions and interests. It is one of the features of science in the Soviet Union, particularly within the Academy, that these varying opinions and interests are more obvious and open than in most other sections of society. Rather than criticising, as some have done, the pluralist approach for claiming a monolithism in Soviet groups that does not exist,[8] we can in fact praise it for putting the emphasis on that lack of monolithism. To understand the role of groups in the Soviet political system we must have some comprehension of the all too often obscure but nevertheless undoubtedly present political process within those groups.

Secondly, it is worth mentioning again, as a mild criticism of institutional pluralism, the importance of personal and non-institutionalised links between scientists and political leaders. These can be based on personal friendship and admiration, one-off approaches from either side, or even pure chance. Some writers on Western science management have suggested that scientists are able to use, and indeed deliberately take special advantage of a non-bureaucratic 'ethos' to avoid formal channels of communication.[9] Perhaps this exaggerates the uniqueness of scientists. Any group or individual able to do so in any society will use informal channels of communication with the political leadership. Certainly one gets a sense from the Soviet and emigre literature of the existence of a self-conscious group at the top of society made up of leading members of many different professions, occupations and social and economic sectors, whether one calls it the *nomenklatura*, new ruling class, partocracy or whatever. But while these people clearly share an interest in maintaining society in essentially its present form, it would be dangerous to put too much emphasis on the identity of their interests within the political process.

While I am not prepared to claim that informal contacts between scientists and political leaders are unique, nor do I see them, as Hodnett has claimed, as a sign of weakness, a compensation for lack of bureaucratic clout.[10] Rather they are additional to any bureaucratic clout scientific institutions might have. We do not have strong evidence (in the natural sciences) that such contacts continue to exist in recent times. Perhaps scientists have lost their status; perhaps in a more settled and institutionalised system they are no longer necessary. My guess is that they still exist, and eventually memoir literature might reveal them.

To sum up the place of scientists in the Soviet policy-making process, one does not find that they dominate it or that they can impose their will on the political leadership. One would not expect to find such power in any pluralist system. But their role in the policy-making process is based on firmer foundations than the purely capricious goodwill of the party leadership, and to that extent has a pluralist character.

One consequence of the emphasis of pluralist theory on inputs has been a neglect of outputs, not so much of the policies that are arrived at, but of their implementation. It is perhaps a consequence of the transfer of pluralism from the Western tradition, which holds, probably inaccurately, that once a policy is arrived at through a democratically pluralist process, all groups, no matter how grudgingly, will accept the new policy and work for its implementation. This certainly is not a feature of modern Soviet society. The ability of powerful bureaucratic interests to frustrate the implementation of policies is perhaps stronger evidence of their 'pluralist' character than their contribution to the making of those policies. This study has not concentrated on such issues, but the difficulties the party has in ensuring the implementation of its policies nevertheless show through strongly.

Another problem with the pluralist model is that so few of the major subfields of Western pluralism are transferable to the Soviet case. Elections, political advertising, public pressure campaigns, etc – the core of Western pluralist studies – cannot be applied to the Soviet Union. Attempts to do so have produced very unconvincing results, viz. many studies of elections and popular participation in the Soviet Union. Given the closed, centralised and bureaucratic nature of the system it is not surprising that Soviet-style pluralism operates within a bureaucratic form of politics, based on formal membership of bureaus and committees and on informal links with and between officials. Therefore, to be used as a practical tool in the examination of the day-to-day management of Soviet society Soviet pluralism must move close to some of the 'vanguard party' models with their stress on organisational behaviour. The theory of organisational behaviour is a field in which there is an extremely rich Western literature, only some of it firmly in the pluralist tradition. Much of it would seem to be a valuable source of ideas and working hypotheses in the study of the Soviet system.

It is to be hoped that such a bureaucratic orientation would protect us from criticism on the grounds that the use of the word pluralism suggests acceptance of the view that democracy of some sort exists in the Soviet Union. There seems little danger of confusing pluralism and democracy in the Soviet case. There is something inherently anti-democratic in Soviet-style pluralism. It is a pluralism based almost entirely on the special and narrowly-defined place of a rather limited number of groups in society. The result is the rigid, inflexible system we find in the Soviet Union.

I do not claim the pluralist approach to be the ideal or last word in Sovietological theory. We do have to overcome the temptation to ignore the oppressive political atmosphere in which pluralist politics takes place. It also has the major shortcoming of not in itself explaining the special place of the party. It serves well in exposing the fallacy in the other approaches that the special place of the party is or must be the same as a totally dominant one. But what we really need is a model which both allows for and measures the party's special role and describes and explains the limitations of that role. As yet we have no such model.

The Soviet Communist Party has most impressive resources at its disposal – a universal ideology, a decision-making structure which it dominates, an apparatus found at all levels of the system, members working in all Soviet collectives, and a proven willingness to use coercive methods. All these resources are applied to the control and management of Soviet science. The most striking finding of this study is not so much that such resources exist, but that they are so dissipated in application through the effects of scientific tradition, the impressive abilities and prestige of some Soviet scientists, the party's lack of expertise, and the opposition and obstructionism of a whole range of sectional interests. I make no claims of a 'free' Soviet science or a powerless party. But if the party uses its resources in other parts of the system as inefficiently as it does in the management of science and technology, some serious thought has to be given to a reevaluation of both its 'leading role' and its totalitarian power.

Notes and References

INTRODUCTION

1. C. J. Friedrich and Z. K. Brzezinski, *Totalitarian Dictatorship and Autocracy* (Praeger, New York, 1966) 2nd edition, p. 21. For a review of the totalitarian model, see T. H. Rigby, ' "Totalitarianism" and change in communist systems', *Comparative Politics*, vol. 4, no. 3 (April 1972) pp. 433–53.
2. L. Schapiro, *The Communist Party of the Soviet Union* (Eyre and Spottiswoode, London, 1970) 2nd edition, pp. 619–29; B. Meissner, 'The Soviet Union and its area of hegemony and influence', in H. J. Veen et al. *Is Communism Changing?* (Hase & Koehler, Mainz, 1981) pp. 27–50.
3. D. Nelidov, 'Ideokraticheskoe soznanie i lichnost', in *Samosoznanie. Sbornik statei* (Khronika, New York, 1976) pp. 117–51.
4. For an example of such an approach, see P. N. Wrinch, 'Science and politics in the USSR: the genetics debate', *World Politics*, vol. 3, no. 4 (July 1951) pp. 486–519.
5. For a brief but devastating critique of the ideological view, see D. Joravsky, 'Bosses and scientists', *Problems of Communism*, vol. 16, no. 1 (January–February 1967) pp. 72–75.
6. A. Shtromas, *Political Change and Social Development. The case of the Soviet Union* (Lang, Frankfurt/Main and Bern, 1981) p. 63.
7. M. Popovsky, *Science in Chains. The crisis of science and scientists in the Soviet Union today* (Collins and Harvill, London, 1980). See also W. D. Connor, 'Differentiation, integration, and political dissent in the USSR', in R. L. Tokes (ed.), *Dissent in the USSR. Politics, ideology and people* (Johns Hopkins UP, Baltimore and London, 1975) ch. 4.
8. For example Popovsky, op. cit.; A. Nekrich, *Otreshis' ot strakha. Vospominaniia istorika* (Overseas Publications Interchange, London, 1979); M. Ya. Azbel, *Refusenik. Trapped in the Soviet Union* (Hamish Hamilton, London, 1981); Z. A. Medvedev, *Mezhdunarodnoe sotrudnichestvo uchenykh i natsional'nye granitsy. Taina perepiski okhraniaetsia zakonom* (Macmillan, London and Basingstoke, 1972); *Soviet Science* (Oxford UP, 1979).
9. For example, C. Zirkle, *Death of a Science in Russia* (University of Philadelphia Press, 1949).
10. V. Turchin, *The Inertia of Fear and the Scientific World View* (Martin Robertson, Oxford, 1981) p. 9.

11. Shtromas, op. cit.
12. T. H. Rigby, 'A conceptual approach to authority, power and policy in the Soviet Union', in T. H. Rigby, A. Brown and P. Reddaway (eds), *Authority, Power and Policy in the USSR. Essays dedicated to Leonard Schapiro* (Macmillan, London and Basingstoke, 1983) ch. 2.
13. F. J. Fleron, 'Co-optation as a mechanism of adaption to change: the Soviet political leadership system', in R. E. Kanet (ed.), *The Behavioral Revolution and Communist Studies* (Free Press, New York, 1971) ch. 5.
14. A. G. Meyer, 'USSR, Incorporated', in D. W. Treadgold (ed.), *The Development of the USSR. An exchange of views* (University of Washington Press, Seattle, 1964) pp. 21–8. See also A. G. Meyer. *The Soviet Political System: an interpretation*, (Random House, New York. 1965).
15. Rigby, 1972, op cit., p. 451.
16. T. H. Rigby, 'Politics in the mono-organizational society', in A. C. Janos (ed.), *Authoritarian Politics in Communist Europe. Uniformity and diversity in one-party states* (Institute of International Studies, University of California, Berkeley, 1976) p. 31.
17. P. Schmitter, 'Still the century of corporatism?', *Review of Politics* vol. 36, no. 1 (January 1974) pp. 93–4.
18. V. Bunce and J. M. Echols III, 'Soviet politics in the Brezhnev era: "pluralism" or "corporatism"?' in D. R. Kelley (ed), *Soviet Politics in the Brezhnev Era* (Praeger, New York, 1980) ch. 1; V. Bunce, 'The Political economy of the Brezhnev era. The rise and fall of corporatism', *British Journal of Political Science*, vol. 13, pt 2 (April 1983) pp. 129–58. See also G. Lembruch, 'Interest intermediation in capitalist and socialist systems. Some structural and functional perspectives in comparative research', *International Political Science Review*, vol. 4, no. 2 (1983) pp. 153–72, and J. F. Hough, 'Pluralism, corporatism and the Soviet Union', in S. G. Solomon (ed.), *Pluralism in the Soviet Union. Essays in honour of H. Gordon Skilling* (Macmillan, London and Basingstoke, 1983) ch. 3.
19. See Article 6 of the 1977 Constitution and the Preamble to the Party Rules.
20. Sectional interests will be selfish and competing until the stage of communism is reached, when there will no longer be scarcity. V. V. Kosolapov and A. N. Shcherban', *Optimizatsiia nauchno-issledovatel'skoi deiatel'nosti* ('Naukova dumka', Kiev, 1971) p. 110.
21. *Sovetskoe gosudarstvo i pravo*, 10/71, p. 91.
22. The main discussion of interest theory took place in the early 1970s, principally in the journal *Ekonomicheskie nauki*. The discussion was brought to an end in 1972, but in friendly and positive terms, and it would seem that the categories analysed in the discussion and their practical implications are considered to be still valid. For the article concluding the discussion, see *Ekonomicheskie nauki*, 5/72, pp. 14–17. See also my unpublished paper, 'A Soviet interest system', Department of Political Science, RSSS, Australian National University, Canberra, April 1974, and R. J. Hill, *Soviet Politics, Political Science and Reform* (Martin Robertson, Oxford, 1980) pp. 85–94.

23. *Kommunist*, 3/83, pp. 13–14.
24. Rigby essentially accepts the party's own characterization of its role in 'The CPSU from Stalin to Chernenko: membership and leadership', paper presented to a conference in honour of Lloyd Churchward, 'Forty Years of Soviet Studies', Melbourne, October 1984, p. 1.
25. Rigby in Janos, op. cit., p. 31.
26. Barrington Moore, *Terror and Progress USSR. Some sources of change and stability in the Soviet dictatorship* (Harvard UP, Cambridge, Mass., 1954) pp. 188–9.
27. For some general accounts of technocracy, see J. Meynaud, *Technocracy* (Faber, London, 1968); J. Ellul, *The Technological Society* (Cape, London, 1965).
28. See J. F. Hough. 'The bureaucratic model and the nature of the Soviet system', *Journal of Comparative Administration*, vol. 5, no. 2 (August 1973) p. 140. For some specific and detailed studies of the technocratic challenge to the party, see A. Parry, *The New Class Divided. Science and technology versus communism* (Macmillan, New York 1966) M. Garder, *Die Agonie des Sowjetsystems* (Ullstein, 1965); Shtromas, op. cit.
29. See below, pp. 31–2
30. Robert McNamara, as US Secretary for Defense, expressed the matter well: 'The possibilities that welled up out of the technological program and the ideas and proposals put forth by the technologists eventually created a set of options that was so narrow in the scope of its alternatives and so strong in its thrust that the political decisions makers had no real independent choice in the matter.' Quoted in H. F. York, *The Advisors. Oppenheimer, Teller and the Superbomb* (Freeman, San Francisco, 1976) p. 11.
31. M. P. Gehlen, 'Group theory and the study of Soviet politics', in S. I. Ploss (ed.), *The Soviet Political Process. Aims, techniques, and examples of analysis* (Ginn, Waltham, Mass., 1971) pp. 40–1.
32. R. J. Hill. T. Dunmore, K. Dawisha, 'The USSR. The revolution reversed?', in L. Holmes (ed.), *The Withering Away of the State? Party and state under communism* (Sage, London and Beverly Hills, 1981) pp. 202–3.
33. J. F. Hough, 'The party apparatchiki', in H. G. Skilling and F. Griffiths (eds), *Interest Groups in Soviet Politics* (Princeton UP, 1971) ch. 3.
34. J. J. Schwartz and W. R. Keech, 'Group influence and the policy process in the Soviet Union', *American Political Science Review*, vol. 62, no. 3 (September 1968) pp. 840–51.
35. T. Gustafson, 'Environmental conflict in the USSR', in D. Nelkin (ed.). *Controversy. Politics of technical decisions* (Sage, London and Beverly Hills, 1979) ch. 3.
36. P. H. Solomon, *Soviet Criminologists and Criminal Policy. Specialists in policy-making* (Macmillan, London and Basingstoke, 1978).
37. J. Lowenhardt, *Decision Making in Soviet Politics* (Macmillan, London and Basingstoke, 1981).
38. For example, T. Dunmore, *The Stalinist Command Economy. The Soviet state apparatus and economic policy, 1945-53* (Macmillan, London and

Basingstoke, 1980); W. O. McCagg, *Stalin Embattled, 1943–1948* (Wayne State UP, Detroit, 1978). See also D. E. Langsam and D. W. Paul, 'Soviet politics and the group approach: a conceptual note', *Slavic Review*, vol. 31, no. 1 (March 1972) p. 137; Shtromas, op. cit., pp. 68–9.

39.   See the pioneering works written and edited by H. Gordon Skilling. H. G. Skilling, 'Soviet and communist politics: a comparative approach', *Journal of Politics*, vol. 22, no. 2 (May 1960) pp. 300–13; Skilling and Griffiths, op. cit.

40.   See, for example, Lembruch, op. cit., p. 155.

41.   J. F. Hough. 'The Soviet system. Petrification or pluralism?', *Problems of Communism*, no. 2 (March–April 1972) pp. 27–8. For support for his view, see E. Jones, 'Committee decision making in the Soviet Union', *World Politics*, vol 36, no. 2 (January 1984) p. 165; Connor in Tokes, op cit., p. 149.

42.   See W. E. Odom, 'A dissenting view on the group approach to Soviet politics', *World Politics*, vol. 28, no. 4 (July 1976) p. 555.

43.   For example, F. J. Fleron, 'Representation of career types in Soviet political leadership', in R. B. Farrell (ed.) *Political Leadership in Eastern Europe and the Soviet Union* (Aldine, Chicago, 1970) pp. 117–20.

44.   See L. L. Lubrano, 'Soviet science specialists: professional roles and policy involvement', in R. B. Remnek (ed.), *Social Scientists and Policy Making in the USSR* (Praeger, New York, 1977) pp. 60–1.

CHAPTER 1

1.   For a detailed account of this debate, see D. Joravsky, *Soviet Marxism and Natural Science, 1917–1932* (Columbia UP, New York, 1961).

2.   As Alexander Vucinich puts it: 'Marxist officialdom was willing to admit that historical materialism did not rest on absolute principles, impervious to the vicissitudes of the human condition; but, at the same time, it reserved for itself alone the right to make theoretical changes, A. Vucinich, *Empire of Knowledge. The Academy of Sciences of the USSR (1917–1970)* (University of California Press, Berkeley and LA, 1984) p. 169.

3.   See Z. A. Medvedev, *The Rise and Fall of T. D. Lysenko* (Columbia UP, New York and London, 1969); D. Joravsky. *The Lysenko Affair* (Harvard UP, Cambridge, Mass., 1970).

4.   See Barrington Moore, op. cit., ch. 5; L. R. Graham, 'Quantum mechanics and dialectical materialism', *Slavic Review*, vol. 25, no. 3 (Spring 1966) pp. 381–410; D. Holloway, 'Innovation in science – the case of cybernetics in the Soviet Union', *Science Studies*, vol. 4, no. 4 (October 1974), pp. 299–337: I. Iakhot, 'Sud'ba statistiki i materialisticheskikh metodov v sotsiologii (30-e gody)', *SSSR. Vnutrennye protivorechiia*, no. 1 (1981) pp. 158–200.

5.   This is the modern Soviet view, when such things are discussed in any detail. See S. R. Mikulinsky and R. Rikhta (eds), *Sotsializm i nauka* ('Nauka', Moscow, 1981) p. 109; V. and V. Dorofeev, *Vremia, uchenye, sversheniia* (Politizdat, Moscow, 1975) p. 93.

6. Marx discussed these matters most explicitly in the *Grundrisse*. The relevant section was published in the Soviet Union only in 1939. It was not used by scholars discussing the nature of science until the late 1950s. P. R. Josephson, 'Science and ideology in the Soviet Union: the transformation of science into a direct productive force', *Soviet Union/ Union Soviétique*, vol. 8, part 2 (1981) pp. 160–2.

7. See his letter to H. Starkenburg, cited in T. H. Rigby and R. F. Miller, *Political and Administrative Aspects of the Scientific and Technical Revolution in the USSR*. Occasional paper no. 11, Department of Political Science, RSSS, Australian National University, Canberra, 1976, pp. 64–5; and the quotation from him in *Marksistsko–leninskoe obrazovanie i nauchnoe tvorchestvo* ('Nauka i tekhnika', Minsk, 1982) pp. 8–9; Joravsky (1961) op. cit., pp. 9–10.

8. Josephson suggests that he did this accidentally in his concern to attack the linguistic theories of Marr. However, presumably the attack on Marr was part of a larger concern to put in their place some of the more objectionable 'Bolshevizers', and perhaps a genuine desire to improve the status of science and reward the scientists who had built the bomb. Josephson, op. cit., pp. 173–4.

9. A. V. Bakunin (ed.). *Deiatel'nost' KPSS po uskoreniiu nauchno-tekhnicheskogo progressa* ('Vysshaia shkola', Moscow, 1980) p. 139; A. A. Smolkina, *Deiatel'nost' KPSS po vosstanovleniiu i razvitiiu nauchno-tekhnicheskogo potentsiala Leningrada (1945–1965 gg.)* (Izd-vo Leningradskogo universiteta, Leningrad, 1983) pp. 80–1. For the speech, see N. A. Bulganin. *O zadachakh po dal'neishemu pod''emu promyshlennosti'tekhnicheskomu progressu i uluchsheniiu organizatsii proizvodstva* (Politizdat, Moscow, 1955).

10. Quoted in Josephson, op. cit., p. 181.

11. M. Lavigne, 'Advanced socialist society', *Economy and Society*, vol. 7, no. 4 (November 1978) pp 372–3; A. Evans, 'Developed socialism in Soviet ideology', *Soviet Studies*, vol 29, no. 3 (July 1977) PP. 419–20.

12. The Bukharinist idea that science should be funded at a greater rate than industrial investment and production was adopted with particular enthusiasm by Mstislav Keldysh, president of the Academy of Sciences throughout the 1960s. Lowenhardt, 1981, op. cit., p. 172; *Vestnik Akademii nauk SSSR (VAN)*, 7/61, p. 23.

13. *XXIII s"ezd Kommunisticheskoi Partii Sovetskogo Soiuza. Stenograficheskyy otchet* (Politizdat, Moscow, 1966) vol. 1, p. 78. Brezhnev again very deliberately described science as having already become a direct productive force in 1973. L. I. Brezhnev. *Leninskim kursom*, vol. 4 (Politizdat, Moscow, 1974), p. 218. At the 24th Party Congress in 1971 he was very ambiguous, declaring that 'the role of science as a direct productive force is being manifested on an increasing scale.' (Quoted in G. W. Breslauer, *Khrushchev and Brezhnev as Leaders. Building authority in Soviet politics* (Allen and Unwin, London, 1982) p. 194.) Other authoritative sources, including *Pravda* editorials and even the resolution of the 1966 Congress, claim only that science is becoming a productive force, albeit at an ever greater pace. I am unsure of the political significance of these differing views, although perhaps

they should be seen in the context of Brezhnev's failure to bring off the Central Committee plenums on science and technology that he called for in 1971 and 1973. B. B. Parrott, *Technology and the Soviet Polity. The problem of industrial innovation, 1928 to 1973*, Ph. D. dissertation, (Columbia University, New York, 1976) pp. 372, 381.

14. B. B. Parrott, *Politics and Technology in the Soviet Union* (MIT Press, Cambridge Mass. and London, 1983) p. 206.

15. Western commentaries on the STR include J. M. Cooper, 'The scientific and technical revolution in Soviet theory', in F. J. Fleron (ed.), *Technology and Communist Culture. The socio-cultural impact of technology under socialism* (Praeger, New York, 1977) ch. 3; 'The scientific and technical revolution in the USSR', paper presented at the NASEES Annual Conference, Cambridge, 1981; E. P. Hoffman, 'Soviet views of "the STR" ', *World Politics*, vol. 30, no. 14 (July 1978) pp. 615–44; R. F. Miller, 'The scientific–technical revolution and the Soviet administrative debate', in P. Cocks, R. V. Daniels, N. W. Heer (eds), *The Dynamics of Soviet Politics* (Harvard UP, Cambridge Mass. and London, 1976) ch. 8. Two recent monographs on specialized aspects of the STR are G. B. Smith, P. B. Maggs, G. Ginsburgs (eds), *Soviet and East European Law and the Scientific-Technical Revolution* (Pergamon, New York, 1981); E. P. Hoffman and R. F. Laird, *The Scientific-Technological Revolution and Soviet Foreign Policy* (Pergamon, New York, 1982).

16. The first public use of the term was apparently made by Bulganin in his 1955 speech, while it was used by Khrushchev to justify some of his reorganisations. Hoffmann, 1978, op. cit., p. 634; Cooper in Fleron, op. cit., p. 153. R. F. Miller mentions the importance of the translation into Russian in 1956 of J. B. Bernal's book, *Science in History*, for drawing the attention of Soviet theorists to scientific and technical revolutions. R. F. Miller 'Soviet policy on science and technology and East-West relations', *World Review*, vol. 19, no. 2 (June 1980) p. 64.

17. See Cooper in Fleron, op. cit., pp. 153–9.

18. G. M. Dobrov, *Nauka o nauke* ('Naukova dumka', Kiev, 1970).

19. For a description of this development in terms of the history of *naukovedenie*, see Y. M. Rabkin, 'The study of science', *Survey*, vol. 23, no. 1 (102) (Winter 1977–8), pp. 134–45.

20. Cooper in Fleron, op. cit., p. 159.

21. For some of the more fantastic, see Cooper (1981) op. cit.

22. Both Andropov and Chernenko referred to it at the June 1983 Central Committee plenum. *Pravda*, 15 June 1983, p. 1; 16 June 1983, p. 1. See also the reference to it in the August 1983 Central Committee and Council of Ministers decree on science and technology. *Pravda*, 28 August 1983, p. 1.

23. *Ideino-politicheskoe vospitanie nauchno-tekhnicheskoi intelligentsii* ('Nauka', Moscow, 1982) p. 32.

24. Medvedev (1969) op. cit., pp. 221–3.

25. *Plenum Tsentral'nogo Komiteta KPSS 24–26 marta 1965 goda. Stenograficheskiy otchet.* (Politizdat, Moscow, 1965) pp. 25–6. Quoted in T. Gustafson, *Reform in Soviet Politics. Lessons of recent policies on land and water* (Cambridge UP, New York, 1981) p. 31.

26. *Kommunist*, 3/67, p. 10.
27. *Novy mir*, 11/77, pp. 175–8; *VAN*, 4/81, p. 28.
28. *New Scientist*, 12 June 1980, pp. 234–7; *Nature*, 14 August 1980, p. 652; 5 February 1981, p. 438.
29. *Literaturnaia gazeta*, 3 June 1981, p. 13.
30. Peive is described as a 'fixist' in one Western publication. However, Soviet sources indicate that he is a 'mobilist'. *New Scientist*, 12 June 1980, p. 236; *Literaturnaia gazeta*, 3 June 1981, p. 13; Iu. M. Pushcharovsky and A. L. Ianshin, *Tektonicheskoe razvitie zemnoi kory i razlomy* ('Nauka', Moscow, 1979) p.5.
31. P. N. Kropotkin, 'A criticism of some tectonic theories of fixism', *Izvestiia AN SSSR. Ser. geologii*, 6/64, p. 12.
32. *VAN*, 7/72, pp. 65–89; 3/73, pp. 49–56; 5/79, pp. 79–83; 12/81, pp. 90–4; 3/82, pp. 110–14. See also the two articles on tectonics in the 1975 edition of *Bol'shaia Sovetskaia Entsiklopediia*.
33. *VAN*, 1/68, pp. 111–12; *Nature*, 30 October 1980, p. 774. Another 'fixist', M. V. Muratov, is chairman of the Interbranch Tectonics Committee. *VAN*, 8/73, p. 117.
34. *New Scientist*, 12 June 1980, p. 236. For results of drilling programmes directed by Belousov that might seem to support his views, see *Nature*, 30 October 1980, p. 774.
35. Indeed one acquaintance in Moscow who is close to the debate, although not an impartial observer, claims that the 'mobilists', particularly Khain, are using unfair methods to repress the views of the 'fixists'. Khain has an advantage over Belousov in that he is a party member, while Belousov is not.
36. M. B. Adams, 'Science, ideology, and structure: the Kol'tsov institute, 1900–1970', in L. L. Lubrano and S. G. Solomon (eds), *The Social Context of Soviet Science* (Westview, Boulder and Davison, Folkestone, 1980) p. 196; *Komsomol'skaia pravda*, 27 February 1965, p. 4.
37. Adams, op. cit; R. Berg, *Sukhovei. Vospominaniia genetika* (Chalidze, New York, 1983) pp. 46–7.
38. Berg, op. cit., esp. pp. 184, 231–2, 245. See also M. Popovsky, *Upravliaemaia nauka* (Overseas Publications Interchange, London, 1978) pp. 272–5
39. *Kommunist*, 11/80, pp. 62–74.
40. See his involvement in the blackballing of Gatovsky for election as a full member of the Academy of Sciences, and his attacks on *IKSI*. A. Katsenelinboigen, 'Nuzhny li v SSSR Don Kikhoty?' *SSSR. Vnutrennye protivorechiia*, no. 2, 1981, pp. 316–17; *VAN*, 7/72, pp.55–65. See also L. R. Graham, *Science and Philosophy in the Soviet Union* (Allen Lane, London, 1973) pp. 122–7.
41. See Berg's sympathetic treatment of him. Op. cit., pp. 236, 266.
42. *VAN*, 6/81, pp.42–7; S. Voronitsyn, 'Academicians clash over genetics', *Radio Liberty*, RL 284/81, 20 July 1981.
43. *VAN*, 12/81, p. 123.
44. *VAN*, 12/81, p. 53.
45. Berg claims that following his dismissal the Central Committee Science Department also dispensed with his services as a consultant on biological matters. Berg, op. cit., p. 249.

46. L. R. Graham, 'Biomedicine and the politics of science in the USSR', *Soviet Union/Union Sovietique*, vol. 8, pt. 2 (1981) p. 151.
47. See the views of Petr Fedoseev, the Academy's chief ideologist, in *Novy mir*, 11/77, pp. 160–83.
48. Presumably it was not coincidental that at this time reports began to appear of a revival of Soviet interest in IQ testing. *Nature*, 25 March 1982, p. 280.
49. *Plenum Tsentral'nogo Komiteta KPSS 14–15 iunia 1983 goda. Stenografichesky otchet* (Politizdat, Moscow, 1983) p. 13.
50. See Fedoseev, *VAN*, 4/83, p. 120. For an interesting account of the scientists' fears of political interference in genetic engineering, see Graham, 1981, op. cit.
51. *VAN*, 11/55, p. 108–111, 10/56, pp. 86–7; 10/59, pp. 95–6.
52. *VAN*, 12/61, p. 15. See Fedoseev writing in 1981 on Keldysh's insistence in the early 1960s that mathematical methods be applied to economics. *VAN*, 2/81, p. 50.
53. Fedoseev claims that *TsEMI* was set up on the initiative of Keldysh. This seems slightly improbable, although his support was undoubtedly important. Ibid.
54. *VAN*, 6/63, pp. 95–6; 3/65, p. 7. See Kantorovich's veiled attack on the traditionalists in 1964. He stated that Soviet scientists must rebuff foreign attempts to turn economics from mathematical methods by claiming that such methods constitute revision of Marxist principles. *VAN*, 4/64, p. 77.
55. Rumiantsev was formerly head of the Science Department of the Central Committee and then consecutively chief editor of *Kommunist, Problemy mira i sotsializma* and *Pravda*. For evidence of 'liberalism' in earlier stages of his career, see Lowenhardt, op. cit., p. 49; G. L. Rocca, ' "A second party in our midst": the history of the Soviet Scientific Forecasting Association', *Social Studies of Science*, vol. 11, no. 2 (May 1981) p. 236. For his still earlier career, see I. Birman, 'Protivorechivye protivorechiia (zametki o sovetskoi ekonomicheskoi teorii i praktiki)', *SSSR. Vnutrennye protivorechiia*, no. 1 (1981) p. 26; W. G. Hahn, *Postwar Soviet Politics. The fall of Zhdanov and the defeat of moderation* (Cornell UP, Ithaca and London, 1982) pp. 152–5.
56. *Politichesky dnevnik. 1964–70* (Fond im. Gertsena, Amsterdam, 1972) p. 235; E. Belitsky, *The Central Institute of Mathematical Economics*, Soviet Institutions Series No. 1, Hebrew University of Jerusalem (September 1974) pp. 40–1.
57. Belitsky, op. cit., p. 29; *VAN*, 2/66, pp. 86–7.
58. Katsenelinboigen claims that Khachaturov was a supporter of mathematical economics as a scientific technique but opposed Fedorenko and *TsEMI* for reasons of personal rivalry. A. Katsenelinboigen, *Soviet Economic Thought and Political Power in the USSR* (Pergamon, New York, 1980), pp. 68–9).
59. M. Ellman, *Soviet Planning Today. Proposals for an optimally functioning economic system* (Cambridge UP, 1971); R. F. Vidmer, 'Soviet studies of organization and management: a "jungle" of competing views', *Slavic Review*, vol. 40, no. 3 (Fall 1981) p. 418.

60. I. Zemtsov, *IKSI. The Moscow Institute of Applied Social Research. A note on the development of sociology in the USSR*, Soviet Institutions Series No. 6, Hebrew University of Jerusalem (April 1976) p. 3. Konstantinov had been well disposed towards sociology since the 1930s. Vucinich (1984) op. cit. p. 167.

61. The first went to the International Sociological Association conference in Holland in 1956. *VAN*, 12/56, pp. 55–61.

62. *VAN*, 6/58, p. 100.

63. *VAN*, 7/61, p. 99; 12/62, p. 54.

64. *VAN*, 1/69, p. 12.

65. Zemtsov, op. cit., pp. 30–1; *VAN*, 9/68, p. 155.

66. Zemtsov, op. cit., p. 36.

67. W. G. Hahn, op. cit., p. 170.

68. For attacks on Levada, see *Vestnik Moskovskogo universiteta. Seriia filosofii*, 3/70, pp. 95–6; *Filosofskie nauki*, 3/70, pp. 178–85.

69. Zemtsov, op. cit.,, pp. 49–58. See also J. W. Hahn, 'The role of Soviet sociologists in the making of social policy', in Remnek, op. cit., p. 40.

70. *VAN*, 4/72, p. 142.

71. *International Herald Tribune*, 13 October 1976, p. 4.

72. *VAN*, 2/72, pp. 3–5. For the Academy's reaction, see *VAN*, 6/72, pp. 17–19; 7/72, p. 9.

73. Belitsky, op. cit., pp. 40–2.

74. *International Herald Tribune*, 13 October 1976, p. 4.

75. *VAN*, 10/72, p. 9; 12/73, p. 54.

76. *VAN*, 2/75, p. 128; 3/75, p. 139; 2/78, pp. 42–7.

77. *VAN*, 11/76, pp. 16–19.

78. *Ideino-politicheskoe vospitanie*, op. cit., p. 9; *Partiinaia zhizn'*, 21/76, p. 26; 5/78, pp. 15–27; *Kommunist*, 12/83, p. 124.

79. *Pravda*, 15 June 1983, p. 1.

80. *VAN*, 12/83, pp. 3–20. See also the criticisms at the meeting of the Academy's Social Sciences Sector. *VAN* 3/84, pp. 42–6, 57.

81. See A. Helgeson, 'Recent developments on the Soviet sociological scene', *Radio Liberty*, RL 461/83, 8 December 1983, p. 4. For earlier calls for such a centre, see *Partiinaia zhizn'*, 9/83, p. 83.

82. *Pravda*, 24 February 1984, p. 1.

83. See Khachaturov in *VAN*, 12/83, pp. 10–11.

84. The degree to which the Soviet leadership has come to use policy-relevant social sciences has been the subject of much analysis in the Western literature in recent years. For example, Gustafson (1981) op. cit; Remnek, op. cit; P. H. Solomon (1978) op. cit.

85. Graham (1973) op. cit.

86. For a brief summary of Marx's position on the place of science, see Josephson, op. cit., pp. 159–62.

87. This is best illustrated by the selection from his *Collected Works* published in *V. I. Lenin, KPSS o razvitii nauki* (Politizdat, Moscow, 1981) pt. 1.

88. L. R. Graham, 'The development of science policy in the Soviet Union', in T. Dixon Long and C. Wright (eds), *Science Policies of Industrial Nations* (Praeger, New York, 1975) p. 12.

89. Miller (1980) op. cit., p. 73.
90. Parrott (1976) op. cit., p. 102; H. D. Balzer, 'The Soviet scientific and technical intelligentsia', *Problems of Communism*, vol. 31, no. 3 (May–June 1982) p. 66.
91. See below, pp. 59–60.
92. See below, pp. 91–95, 106–110.
93. See Rigby and Miller, op. cit., p. 64.
94. See Parrott (1976) op cit., pp. 126–58.
95. See R. Lewis, *Science and Industrialization in the USSR. Industrial research and development, 1917–1940* (Macmillan, London and Basingstoke, 1979) ch. 8.
96. For example, *VAN*, 11/80, p. 63; 9/82, p. 21; *Pravda*, 26 February 1981, p. 3.
97. See his statement in the *Economic and Philosophical Manuscripts* (1844): 'Historiography itself only takes natural science incidentally into account, regarding it as a factor making for enlightenment, for practical utility, and for particular great discoveries. But the natural sciences have penetrated all the more *practically* into human life, through their transformation of industry. They have prepared the emancipation of humanity, even though their immediate effect may have been to accentuate the dehumanizing of man. *Industry* is the real historical relation of Nature, and thus of the natural sciences, to man.' T. B. Bottomore and M. Rubel (eds), *Karl Marx. Selected writings in sociology and social philosophy* (Penguin, Harmondsworth, 1967) p. 8.
98. Vucinich (1984) op. cit., ch. 3.
99. See S. Fortescue, *The Academy Reorganized. The R&D role of the Soviet Academy of Sciences since 1961*, Occasional paper No. 17, Department of Political Science, RSSS, Australian National University, Canberra, 1983, chs 1 and 4.
100. Graham in Long and Wright, op. cit., p. 20.
101. See K. E. Bailes, 'The politics of technology. Stalin and technocratic thinking among Soviet engineers', *American Historical Review*, vol. 79, no. 2 (April 1974) pp. 445–69.
102. See S. Fortescue, 'Party membership in Soviet research institutes', *Soviet Union/Union Sovietique*, vol. 11, pt. 2 (1984) p. 139.
103. Between the 1930s and the 1970s the ratio of wages of engineering and technical personnel to those of workers fell from 2.5:1 to 1.4:1. L. S. Bliakhman and O. I. Shkaratan, *NTR, rabochy klass, intelligentsiia* (Politizdat, Moscow, 1973) p. 264. This book contains considerable data on comparative pay scales and living conditions. Between 1960 and 1977 the average pay of people working in science (this includes ancillary workers, who are generally paid worse if they work in science than in industry) showed one of the slowest increases among all branches of the economy. The highest and lowest increases, in rubles per month, were as follows: art – 46.1; health and social welfare – 49.8, administration – 50.3; science – 53.9; construction – 92.4, transport – 99.2. Over that period science went from being the best paid category to the fourth best, with industrial personnel being one of the groups moving ahead of them. The advantage the average scientific salary had over the national average declined from 37.3 per cent to 6.1 per cent. An improve-

ment in the scientists' situation then became evident, following the 1976 decision to raise pay rates in science and the increasing access of scientists to bonus schemes. Between 1977 and 1980 salaries rose on average 14.9 rubles, as against 13.7 for the economy as a whole. However, science remained in fourth place, and its advantage over the national average had declined to 5.9 per cent. G. A. Lakhtin (ed.), *Osnovy upravleniia naukoi. Ekonomicheskie problemy* ('Nauka', Moscow, 1983) pp. 149–50; *VAN* 6/77, pp. 49–50; V. A. Pokrovsky, *Uskorenie nauchno-tekhnicheskogo progressa. Organizatsiia i metody* ('Ekonomika', Moscow, 1983) p. 85.

104. For a most interesting presentation of the 'social contract' hypothesis, see V. Zaslavsky, *The Neo-Stalinist State. Class, ethnicity and consensus in Soviet society* (Sharpe, New York, 1982). For an example of the resentment such policies can arouse among the intelligentsia, see A. Fedoseev, *Zapadnia. Chelovek i sotsializm* (Posev, Frankfurt/Main, 1976) p. 97.

105. Evans, op. cit., p. 420.

106. See M. Matthews, *Class and Society in Soviet Russia* (Allen Lane, London, 1972) pp. 120–2; M. Yanowitch and N. T. Dodge, 'The social evaluation of occupations in the Soviet Union', *Slavic Review*, vol.28, no. 4 (December 1969) pp. 619–41; R. W. Mouly, 'Values and aspirations of Soviet youth', in Cocks, Daniels, Heer, op. cit., pp. 225–6; D. Lane and F. O'Dell, *The Soviet Industrial Worker. Social class, education and control* (Martin Robertson, Oxford, 1978) pp. 73–7; V. N. Shubkin, 'Sotsiologicheskie problemy vybora professii', and V. V. Vodzinskaia, 'O sotsial'noi obuslovlennosti vybora professii', in G. V. Osipov and J. Szczepanski (eds), *Sotsial'nye problemy truda i proizvodstvo* ('Mysl'' and 'Ksiazka i wiedze', Moscow and Warsaw, 1966) pp. 23–38, 39–61.

107. See *Nature*, 24 January 1980, p. 322; 3 December 1981, p. 392; *VAN*, 3/84, p. 12. Boris Paton, president of the Ukrainian Academy of Sciences, is particularly concerned. *Izvestiia*, 1 January 1984, p. 3; *VAN* 12/81, p. 39; *Komsomol'skaia pravda*, 22 March 1980, pp. 1–3.

108. For this and other examples, see Gustafson in Nelkin, op. cit.

109. For resumes of these two works, see G. Hosking, *Beyond Socialist Realism. Soviet fiction since Ivan Denisovich*, (Granada, London, 1980), pp. 87–92.

110. Ibid., p. 82.

111. Quoted in I. Gvat', 'The creativity of F. M. Dostoevsky as a permanent provocation', *Radio Svoboda*, RS 138/83, 13 July 1983, p. 2.

112. See A. Vucinich, *Science in Russian Culture 1861–1917.* (Stanford UP, 1970) pp. 238–9. See also Vucinich (1984) op. cit., pp. 349–50.

113. Hosking, op. cit., p. 82.

114. S. Yurenen, 'Journal politics from Andropov to Chernenko', *Radio Liberty*, RL 299/84, 7 August 1984, p. 4.

115. Cooper (1981) op. cit., p. 4; Graham (1981) op. cit., pp. 152–3.

116. See, for example, M. Brumas, 'USSR discriminates against Jews in higher education', *Radio Liberty*, RL 301/84, 6 August 1984, p. 2; Zaslavsky, op. cit., p. 17.

117. I use the term 'false science' as a translation of the Russian word *lzhenauka*.

118. E. Ashby, *Scientist in Russia* (Penguin, Harmondsworth, 1947) p. 117.
119. R. M. Mills, 'Lysenkoism: the un-science', *Problems of Communism*, vol. 21, no. 5 (September–October 1972) p. 85. See also A. Vucinich, *The Soviet Academy of Science*, Hoover Institute Studies, Series E: Institutions, No. 3 (Stanford UP, January 1956) pp. 49-53.
120. See below pp. 105–6.
121. *New York Times*, 18 August 1980, p. 17.
122   *VAN*, 4/81, p. 87. Zel'dovich was also supported by Nikolai Blokhin, the president of the Academy of Medical Sciences, on this issue, but not in his defence of Freud, whose techniques retain the status of 'false science' in the USSR. Ibid, p. 28. For more on Dzhuna, see *New York Times*, 28 August 1980, p. 11; *Komsomol'skaia pravda*, 16 August 1980, p. 4; *Literaturnaia gazeta*, 27 August 1980, p. 13.
123. For some examples, see *VAN*, 4/60, p. 92; 6/78, p. 31; 6/83, pp. 34–5. See also the major series of articles on the subject in *Literaturnaia gazeta* throughout 1977.
124. For some Western evaluations of Soviet science, see L. L. Lubrano, 'National and international politics in US-USSR cooperation', *Social Studies of Science*, vol. 11, no. 4 (November 1981) pp. 456–9; *New Scientist*, 10 November 1983, p. 442; *Nature*, 14 October 1982, p. 568; F. Narin, J. Davidson Frame, M. P. Carpenter, 'Highly cited Soviet papers: an exploratory investigation', *Social Studies of Science*, vol. 13, no. 2 (1983) pp. 307–19. For a self-evaluation by Soviet scientists, see D. D. Raikova, 'Osobennosti otnosheniia uchenykh k razlichnym sredstvam nauchnoi kommunikatsii', in A. A. Zvorykin (ed.), *Nauchny kollektiv. Opyt sotsiologicheskogo issledovaniia* ('Nauka', Moscow, 1980) pp. 91–5.
125. For examples from different disciplines, see W. C. Clemens Jr, 'Alexander Yanov – writing between the lines', *Problems of Communism*, vol. 27, no. 3 (May–June 1978) pp. 67–8; A. Besancon, 'Solzhenitsyn at Harvard', *Survey*, vol. 24, no. 1 (106) (Winter 1979) pp. 133–44; T. Gustafson, 'American science and Soviet concerns', *Survey*, vol. 23, no. 1 (102) (Winter 1977–8) p. 158; Smith, Maggs and Ginsburgs, op. cit., p. 15. For an historical view, see A. Vucinich, *Social Thought In Tsarist Russia. The quest for a general science of society, 1861–1917* (University of Chicago Press, Chicago and London, 1976) p. 9.
126. See footnote 124.
127. See P. H. Solomon, 'Soviet policy-making in comparative perspective', *Soviet Union/Union Sovietique*, vol. 8, pt. 1 (1981) p. 118.
128. *The Economist*, 26 July 1984, pp. 69–70; S. Voronitsyn, 'Attempts to reach a compromise on diversion of rivers', *Radio Liberty*, RL 313/84, 20 August 1984.
129. A. Kol'man, *My ne dolzhny byli tak zhit'* (Chalidze, New York, 1982) p. 213.
130. F. Franks, *Polywater* (MIT Press, Cambridge Mass. and London, 1981). See also Franks' account of the Soviets' enthusiasm for magnetized water, pp. 46–7.
131. For a summary of such methods, see R. Conquest, *The Politics of Ideas in the USSR* (Bodley Head, London, 1967), ch. 5.

132. For more on these, see below pp. 126–8.
133. See Kosygin in 1961 and a 1967 *Pravda* editorial. *VAN*, 7/61, p. 99–100; *Pravda*, 16 August 1967, p. 1.
134. Through the work-place 'food parcel' system and *podsobnie* farming enterprises.
135. This is particularly evident in the 1983 Charter of Workers' Collectives. See E. Teague 'The USSR Law on Work Collectives: Workers' Control or Workers Controlled?' *Radio Liberty*, RL 184/84, 10 May 1984; *Russkaia mysl'*, 30 June 1983, pp. 1, 4.
136. For example, take the views of B. A. Frolov, as summarised by Linda Lubrano: 'Frolov asserts that the same motivation (of creativity and personal success) exist among Soviet scientists, but the 'spirit of the scientific collective' (in the USSR) transforms individual ambitions into socially useful creativity. He regards the collective as an essential form of scientific interaction and an impetus to scientific discoveries.' L. L. Lubrano, *Soviet Sociology of Science*, American Association for the Advancement of Slavic Studies (Columbus Ohio, 1976) p. 67. As Lubrano makes clear, not all Soviet sociologists of science are as dogmatic in their evaluations of the collective. However party spokesmen such as Yuri Sviridov and Rudolf Ianovsky clearly share the views of Frolov. See footnote 169.
137. For example, P. Kneen, 'Why natural scientists are a problem for the CPSU', *British Journal of Political Science*, vol. 8, part 2 (April 1978) p. 184; Popovsky (1978) op. cit., pp. 268–9.
138. *Kommunist*, 18/68, p. 43; *Sovetskaia Rossiia*, 28 August 1970, p. 2.
139. For example, *Pravda*, 1 August 1973, p. 3; *Partiinaia zhizn'*, 12/79, pp. 58–61; 7/80, p. 36. See the survey results listed in *Ideino-politicheskoe vospitanie*, op. cit., pp. 48–51.
140. See the figures on p. 127.
141. For example, *Voprosy filosofii*, 10/74, p. 162.
142. For example, *Kommunist*, 18/68, p. 38; *VAN*, 11/59, p. 112; 4/80, pp. 44–5; Bakunin, op. cit., p. 132.
143. Mouly, op. cit., p. 234; L. S. Feuer, 'The intelligentsia in opposition', *Problems of Communism*, vol. 19, no. 6 (November–December 1970) p. 8.
144. Turchin, op. cit., p. 162; A. Amal'rik, *Zapiski dissidenta* (Ardis, Ann Arbor, 1982) p. 68.
145. Kneen (1978) op. cit., p. 190.
146. L. L. Greenberg, 'Soviet science policy and the scientific establishment', *Survey*, vol. 17, no. 4 (81) (Autumn 1971) p. 62.
147. Lubrano in Remnek, op. cit., pp. 72–3; Kneen (1978) op. cit., p. 19. One good example of this was the efforts of physicists to find protected places in which geneticists could work during the Lysenko period. Berg, op. cit., p. 225; Lubrano in Remnek, op. cit., p. 79.
148. See below pp. 61–2.
149. Medvedev (1979) op. cit., pp. 100–2.
150. Iu. Glazov, *Tesnye vrata. Vozrozhdenie russkoi intelligentsii* (Overseas Publications Interchange, London, 1973) p. 158; Berg, op. cit., p. 229.
151. For example, the 1970 letter to Brezhnev signed by prominent intellec-

tuals, including academicians Artsimovich, Sagdeev, Engel'gardt, Migdal, Pontekorvo and Alikhanian, in protest against Kochetov's Stalinist novel *Chego zhe ty khochesh*. *Politichesky dnevnik II. 1965–70* (Fond im Gertsena, Amsterdam, 1975) p. 626.

152. See the isolation of Sakharov, and the experiences of Vladimir Turchin and Mark Azbel. Turchin, op. cit., p. 25; Azbel, op. cit., pp. 365–6.
153. A. Solzhenitsyn, *Bodal'sia telenok s dubom. Ocherki literaturnoi zhizni* (YMCA, Paris, 1975) pp. 249–50; *Politichesky dnevnik, 1964–70*, op. cit., pp. 303–6. See also the 'fading away' of the 'liberal' party committee of the Institute of History in 1967. Nekrich, op. cit., p. 300.
154. Medvedev (1972) op. cit., pp. 499–500, 567–8; Popovsky (1978) op. cit., pp. 223–4; Feuer, op. cit., pp. 3–4; R. Medvedev, *Kniga o sotsialisticheskoi demokratii* (Alexander Herzen Foundation and Editions Grasset & Fasquelle, Amsterdam and Paris, 1972) p. 183; Medvedev, (1979), op. cit., p. 135; *Kommunist*, 18/68, pp. 36–45.
155. See below, Chapter 4.
156. Soviet sources still suggest that bourgeois propaganda has a dangerous hold over some members of the scientific intelligentsia. *Ideinopoliticheskoe vospitanie*, op. cit., pp. 115, 163.
157. *Voprosy filosofii*, 4/69, pp. 55–7. For another Estonian survey giving similar if statistically absurd results (the response rates for most variables are well over 100 per cent), see *Kommunist Estonii*, 3/73, p. 44. See also *Kommunist*, 18/68, p. 43.
158. See a survey of young researchers in the Polzunov Central Steam Turbine Institute in Leningrad; research workers in Novosibirsk; and researchers in four institutes in Irkutsk. *Kommunist*, 18/68, p. 43; Popovsky, (1978), op. cit., pp. 223–4; *Pravda*, 7 June 1969, p. 3. For further discussion of these and other surveys, see Lane and O'Dell, op. cit., pp. 73–9.
159. R. Orlova, *Vospominaniia o neproshedshem vremeni* (Ardis, Ann Arbor, 1983); N. Ianevich, 'Institut mirovoi literatury v 1930-e – 1970-e gody', *Pamiat'. Istorichesky sbornik*, no. 5 (1982), pp. 83–162.
160. It is said that research institutes in Novosibirsk tend to be staffed almost entirely by graduates of a single *vuz*. The rivalry between the *vuzy* leads to rivalry between the institutes. *Sotsial'nye issledovaniia*, no. 30 (1970) p. 181.
161. *Literaturnaia gazeta*, 10 June 1981, p. 10.
162. *Nature*, 18 December 1975, p. 566.
163 *Newsweek*, 20 June 1983, p. 60.
164. Azbel, op. cit., p. 264.
165. *Politichesky dnevnik. 1964–70*, op. cit., pp. 300–4.
166. Cited in L. L. Lubrano, 'Survey research as a source of information for Soviet science policy', *Soviet Union/Union Soviétique*, vol. 9, part 1 (1982) p. 63.
167. *Pravda*, 7 June 1969, p. 3.
168. Cited in Feuer, op. cit., pp. 4–5.
169. As we know, such symptoms are part of the psychiatric disorder 'creeping schizophrenia'. *Novy mir*, 11/77, p. 166. See also *Kommunist*, 18/68, p. 43; *Sovetskaia Rossiia*, 28 August 1970, p. 2; R. G. Ianovsky, *Formirovanie lichnosti uchenogo v usloviiakh razvitogo sotsializma*

('Nauka', Novosibirsk, 1979) pp. 105–6; *Pravda*, 14 May 1966, p. 4. (The author of the last article was previously secretary of the party committee of the Siberian department of the Academy of Sciences and later a member of the Novosibirsk *gorkom*.)

170. No one in the modern age would expect to work entirely independently of an institutional base. What we mean by 'working in a collective' is being bound tightly into collective work within an institution.

171. One survey declares that 6.04 per cent of working time is lost through conflicts at work. D. M. Gvishiani (ed.), *Voprosy teorii i praktiki upravleniia i organizatsiia nauki* ('Nauka', Moscow, 1975) p. 248. Despite their relatively low opinion of working in a collective, Soviet scientists attach a very high value to working in a good collective. They appear to mean by this one free of conflict. Ibid, p. 205. Seventeen per cent of conflict situations in one survey resulted in people leaving their jobs. *Literaturnaia gazeta*, 6 May 1970, p. 12.

172. It is said that anonymous denunciations are particularly common among research workers. J. Wishnevsky, 'Soviet legislation and anonymous denunciations', *Radio Liberty*, RL 98/84, 2 March 1984, pp. 5–6. A Soviet writer comments that scientists face conflict between their desire for an administrative career, which improves their material conditions but spoils relations with their colleagues, and their desire for a scientific career. Kosolapov and Shcherban', op. cit., pp. 133–4. See also Mark Popovsky's comments on the quarrelsomeness of Soviet laboratories. Popovsky (1978) op. cit., pp. 203–4.

173. See Fortescue, *The Academy Reorganized*, op. cit., p. 63. It is not surprising that, as Lubrano reports, *kollektivnost'* is ranked higher among senior researchers than junior researchers, since collective research gives them access to the labour of their junior colleagues. Lubrano (1982) op. cit., pp. 64–66. See also S. A. Kugel', B. D. Lebin, Iu. S. Meleshchenko (eds), *Nauchnye kadry Leningrada. Struktura kadrov i sotsial'nye problemy organizatsii truda* ('Nauka', Leningrad, 1973) p. 119.

174. *Ideino-politicheskoe vospitanie*, op. cit., p. 45.

175. Fortescue, *The Academy Reorganized*, op. cit., pp. 63, 64.

176. Even the more decentralised procedures of recent years, which have removed centrally set personnel levels, are still limited by rigidly and centrally set wages funds and ranges. Some science administrators in the metropolitan cities, particularly Leningrad, complain of strict residence permit rules which prevent them from importing personnel from elsewhere. A 1966 government decree banned the building of new scientific institutions in a number of cities without special permission. V. I. Duzhenkov, *Problemy organizatsii nauki (Regional'nye aspekty)* ('Nauka', Moscow, 1978) p. 46; *VAN*, 9/76, p. 51.

177. The supply of apparatus, isotopes, chemical compounds, etc. is planned, which means that orders have to be submitted as much as two years in advance. In one institute it was reported that between one third and one half of working time was spent on travelling around the country trying to obtain essential supplies. *VAN*, 8/77, pp. 41–8; Gvishiani, op. cit., p. 206.

178. Bakunin, op. cit., p. 145; V. A. Sisnev and T. T. Stepanov (eds), *Narodny kontrol' v sfere nauki, obrazovaniia, kul'tury, zdravookhraneniia* ('Iuridizdat', Moscow, 1980) pp. 53–4; *Ekonomicheskaia gazeta*, no. 51 (December 1966) p. 6; *Izvestiia*, 11 July 1964, p. 3; *Literaturnaia gazeta*, 8 March 1967, p. 11; Kosolapov and Sherban', op. cit., p. 178. To this must be added the amount of time Soviet scientists have to devote to 'ancillary' tasks, due to shortages of research and laboratory assistants. Gvishiani, op. cit., p. 172.

179. See the complaints of academician Kantorovich, a Nobel Prize winner, about having to account for his telephone calls and postage stamps. *VAN*, 3/67, p. 203.

180. One of the best examples of this inability to come to terms with change was the closing down of such entrepreneurial organizations as Novosibirsk's *'Fakel'*. J. Lowenhardt, ' "The tale of the torch". Scientist entrepreneurs in the Soviet Union', *Survey*, vol. 20, no. 4 (93) (Autumn 1974) pp. 113–21.

181. This is evident in the data on conflict situations given above. That even very well-known scientists are not spared the humiliations of the Soviet bureaucratic culture is evident from the experiences of O. Iu. Shmidt, after he had given up his official posts, as related by academician Kolmogorov: 'Otto Iul'evich regarded his "demotion" with humour. Once we went together on institute business to the presidium of the Academy. We sat there on a sofa for several hours, uncertain whether we would ever be received. On leaving he said: "Perhaps in the past I didn't always notice petitioners passing their hours in such a situation." ' Dorofeev, op. cit., p. 99.

182. *VAN*, 6/79, pp. 52–9; 11/75, pp. 15–23, esp. p. 18; 12/78, p. 69; Fortescue, *The Academy Reorganized*, op. cit., pp. 73–78; *Literaturnaia gazeta*, 12 July 1978, p. 11.

183. I ignore here the baleful effects on R&D lead times.

184. See Lubrano (1982), op. cit., p. 69; *VAN*, 7/82, pp. 112–13.

185. One survey shows that job dissatisfaction is greatest among the middle-aged (35–44 years), presumably when the choked promotion ladder becomes most frustrating. *Ideino-politicheskoe vospitanie*, op. cit., p. 81.

186. The phrase is used by Ianovsky in *Sovetskaia Rossiia*, 28 August 1970, p. 2. One Soviet source claims that 'internally motivated' young researchers are more likely to be involved in conflict situations than those who are 'externally motivated'. *Problemy rukovodstva nauchnym kollektivom. Opyt sotsial'no-psikhologicheskogo issledovaniia* ('Nauka', Moscow, 1982) p. 157.

187. *Komsomol'skaia pravda*, 21 October 1973, p. 2.

188. See above pp. 22–4.

189. In particular, Pavlov, Vernadsky, Kapitsa. See Joravsky (1961) op. cit., pp. 71–5; Vucinich (1984) op. cit., esp. p. 182.

190. Berg, op. cit., pp. 226–71, 320–2. For another example of such ambiguity in the values and behaviour of science managers, see Khalatnikov, the director of the Landau Institute of Theoretical Physics. Azbel, op. cit., pp. 261–2. The same could also be said of Keldysh and Aleksandrov,

two presidents of the Academy of Sciences. See above, pp. 48–9.

191. For example, *Russkaia mysl'*, 12 July 1984, p. 6; Nekrich, op. cit. p. 164; Ianevich, op. cit., pp. 83–162; *Literaturnaia gazeta*, 26 May 1976, p. 12.

192. The interesting surveys carried out by Nina Toren of emigre scientists in Israel indicate that Soviet scientists have a higher commitment to the social utility of their work than emigre American scientists, that is, a higher degree of one aspect of *partiinost'*. However they have a higher degree of commitment to the Mertonian value 'universalism' and a higher degree of distrust of the Mertonian version of *kollektivnost'*, 'communalism'. N. Toren, 'Scientific autonomy East and West: a comparison of the perceptions of Soviet and United States scientists', *Human Relations*, vol. 32, no. 8 (1979) pp. 643–57; 'The new code of scientists' *IEEE Transactions on Engineering Management*, 3/80, pp. 78–84.

193. The Soviet expression goes: 'Science is the satisfaction of one's own personal curiosity at the expense of the state.'

194. Fortescue, 'Party membership', op. cit.

195. Bliakhman and Shkaratan, op. cit., pp. 167, 168; Kugel', Lebin, Meleshchenko, op. cit., p. 79.

196. Bliakhman and Shkaratan, op. cit., p. 178; *Sovetskaia sotsiologiia*, vol. 1, Sotsiologicheskaia teoriia i sotsial'naia praktika ('Nauka', Moscow, 1982) pp. 125–6. There is also a more than 50 per cent chance that they will marry someone of the same social stratum. Bliakhman and Shkaratan, op. cit., p. 171. See also Kugel', Lebin, Meleshchenko, op. cit., pp. 79–82; M. Matthews, *Education in the Soviet Union. Policies and institutions since Stalin* (Allen and Unwin, London, 1982) p. 161.

197. Shubkin, op. cit., p. 43. Soviet science is characterised by a number of 'dynasties'. One famous Soviet science family is the Keldyshs: Vsevolod Mikhailovich, a construction engineer; his son Mstislav and daughter, both mathematicians, with the former being president of the Academy of Sciences from 1961 to 1975; Mstislav's nephew, Leonid Veniaminovich, a physicist. Mstislav is related to academician P. S. Novikov by marriage. The Gorbunovs so far make up three generations – the geographer who was permanent secretary of the Academy – the geographer who was permanent secretary of the Academy in the mid-30s and two generations of physicists. See *Sovetskaia Rossiia*, 17 January 1980, p. 2. The father of the present Chief Scientific Secretary of the Academy, G. K. Skriabin, was a famous helminthologist, while his son is a biologist who won an Academy medal for young scientists in 1980. *VAN*, 6/80, p. 139. Mikhail Lavrent'ev's son gained full membership of the Academy in the 1981 elections. Such examples could be repeated many times over.

198. R. Amann, 'The Soviet research and development system. The pressure of academic tradition and rapid industrialization', *Minerva*, vol. 8, no. 2 (April 1970) pp. 222–3. Raisa Berg's memoirs provide some sense of a continuity in the values of the Academy elite. Berg. op. cit. See also Amalrik, op. cit., p. 69.

199. The Moscow mathematical school is one that survived hard times in

the early 1930s relatively unscathed. See Joravsky (1961) op. cit., pp. 242–4; R. Medvedev, *Let History Judge. The origins and consequences of Stalinism* (Macmillan, London, 1972) p. 226.

200. Lubrano (1976) op. cit., pp. 54–5; (1982) op. cit., pp. 69–70.
201. Medvedev (1969) op. cit., p. 31.
202. See, for example, J. W. Koning, *The Scientists Look at Research Management* (AMACOM, New York, 1976).

CHAPTER 2

1. See J. F. Hough and M. Fainsod, *How the Soviet Union is Governed* (Harvard UP, Cambridge Mass., 1979) pp. 466–73; J. Lowenhardt, *The Soviet Politburo* (Canongate, Edinburgh, 1982).

2. Most important decrees concerning science since the death of Stalin are contained in *V. I. Lenin, KPSS o razvitii nauki*, op. cit., pt. 3. One could add to the list: 'On measures for improving the coordination of scientific research work in the nation and the work of the Soviet Academy of Sciences', CC-CofM decree published in *Pravda*, 12 April 1961; 'On the work of the party committee of the Lebedev Physics Institute', CC decree published in *Partiinaia zhizn'*, 21/70, pp. 8–10; 'On the work of the party organization of the Institute of Economics of the Soviet Academy of Sciences in the fulfilment of the CC decree "On measures for the further development of the social sciences and raising their role in communist construction" ', CC decree published in *VAN*, 2/72, pp. 3–5; plus the more recent decrees mentioned below.

3. For example, the buro of the Novosibirsk *obkom* examined the work of the party organisation of the Institute of Physics of the Siberian Academy in the light of the Central Committee decree on the Lebedev Physics Institute. Ianovsky (1979) op. cit., p. 198.

4. *Partiinaia zhizn'*, 19/84, p. 3; *Leningradskaia pravda*, 1 March 1985, p. 1.

5. Politburo meetings reported in *Pravda* on 27 May and 6 August 1983. The Central Committee has previously issued decrees on microbiology in August 1970, April 1976 and July 1981.

6. *Pravda Ukrainy*, 30 November 1982, p. 1.

7. *Pravda*, 20 October 1983; 24 February 1984; *Ekonomicheskaia gazeta*, no. 33 (August 1984) p. 3.

8. In the case of the Siberian Academy an essentially favourable decree was published. *Pravda*, 11 February 1977. No decree ever appeared on the Ukrainian Academy, although references have been made to 'a decision of the secretariat', again in favourable terms. *VAN*, 5/77, pp. 6–7; *Pravda Ukrainy*, 17 March 1977, p. 2. In some sources mention is made of an unnamed decree dated January 1977. Bakunin, op. cit., p. 124. An August 1976 *Pravda* editorial states simply that 'in recent days' the Ukrainian Academy's work has been approved (*odobren*) by the Central Committee. *Pravda*, 9 August 1976, p. 1. We have slightly more detail on the operational involvement of the union republic Central Committee Buros in science management. The Buro of the Belorussian Central Committee in 1983 examined the work of the biology institutes

of the Belorussian Academy of Sciences. It found that they were not doing enough to implement the results of their resesarch in production, and the presidium and party committee of the Academy were directed to take measures. *Sovetskaia Belorussiia*, 13 December 1983, p. 1. The Buro of the Armenian Central Committee in 1980 critically examined the work of the Erevan Research Institute of Mathematical Machines in the production of the Nairi-4 computer, in what was clearly a reaction to a crisis situation. *Voprosy istorii KPSS*, 10/82, p. 28. The Buro of the Uzbek Central Committee has also been reported as examining the work of the Uzbek Academy of Sciences in *vnedrenie*. *Ekonomicheskaia gazeta*, no. 27 (July 1979) p. 7.

9.   *VAN*, 3/81 p. 84.

10   G. D. Komkov, B. V. Levshin, L. K. Semenov, *Akademiia nauk SSSR. Kratky istorichesky ocherk*, vol. 2, 1917–76 ('Nauka', Moscow, 1977) 2nd edition, p. 417.

11.   *KPSS v rezoliutsiiakh i resheniiakh s'ezdov, konferentsii i plenumov TsK*, vol. 11, 1972–75 (Politizdat, Moscow, 1976) pp. 322–5.

12.   *Sovetskaia Rossiia*, 22 March 1985, p. 3.

13.   Cited in G. Guroff, 'The Red-expert debate. Continuities in the state–entrepreneur tension', in G. Guroff and F. V. Carstensen (eds), *Entrepreneurship in Imperial Russia and the Soviet Union* (Princeton UP, 1983) p. 213.

14.   See *V. I. Lenin, KPSS o razvitii nauki*, op. cit., part 1. That Lenin was not always so benign is made clear by Mark Popovsky. Popovsky (1978) op. cit., pp. 20–22. His 1919 letter to Gorky is particularly instructive.

15.   D. Anastasyn and I. Voznesensky, 'Nachalo trekh natsional'nykh akademii', *Pamiat'. Istorichesky sbornik*, vyp. 5, 1982, p. 196.

16.   A. Tuominen, *The Bells of the Kremlin. An experience in communism* (University Press of New England, Hanover and London, 1983) p. 171. For another similar case, see *Khrushchev Remembers. The last testament* (Deutsch, London, 1974) pp. 96–7.

17.   R. Medvedev, *Let History Judge*, op. cit., p. 523.

18.   *Khrushchev Remembers: The Last Testament* op. cit., pp. 58–60, 96. See also B. A. Chotiner, *Khrushchev's Party Reform. Coalition building and institutional innovation* (Greenwood, Westport, Conn., 1984) pp. 21–2.

19.   Kol'man, op. cit., pp. 192–4.

20.   In *Tselina* he claims that one of his first meetings on arriving in Kazakhstan was with D. A. Kunaev, then the president of the Kazakh Academy of Sciences (now Kazakh party first secretary), who recommended the director of the Institute of Soil Science as Brezhnev's agricultural consultant. In his struggle with the Ministry of Agriculture he used the probably not altogether objective findings of 'his' scientists to support his case. L. I. Brezhnev, *Tselina* (Politizdat, Moscow, 1978) pp. 14, 17.

21.   See below pp. 82–3.

22.   See his speech at the April 1985 Central Committee plenum. *Pravda*, 24 April 1985, p. 1.

23.   See B. A. Ruble, 'Romanov's Leningrad', *Problems of Communism*, vol. 32, no. 6. (November–December 1983) pp. 36–48. One can presum-

ably see Romanov's influence, at the time in the Central Committee apparatus, in the August 1984 Central Committee decree praising and recommending the broad application of Leningrad methods of R&D planning and management. *Ekonomicheskaia gazeta*, no. 33 (August 1984) p. 3.

24. For speculation on his role, see J. F. Hough, 'Soviet policy-making toward foreign communists', *Studies in Comparative Communism*, vol. 15, no. 3 (Augumn 1982) p. 177. Anatoly's son in turn seems to be taking up a career in international relations. G. Jozsa, 'Political *Seilschaften* in the USSR', in T. H. Rigby and B. Harasymiw (eds), *Leadership Selection and Patron-Client Relations in the USSR and Yugoslavia* (Allen and Unwin, London, 1983) p. 151; *Novaia i noveishaia istoriia*, 3/80, p. 224.

25. A. Rahr, 'Igor' Andropov to be new Soviet Ambassador to Greece', *Radio Liberty*, RL 281/84, 19 July 1984.

26. L. Vladimirov, *Sovetsky kosmichesky blef* (Posev, Frankfurt/Main, 1973) p. 56. Gallagher and Spielmann claim that Khrushchev showed favour to Chelomei in order to goad Korolev on to work harder. M. P. Gallagher and K. F. Spielmann, *Soviet Decision-Making for Defense. A critique of US perspectives on the arms race* (Praeger, New York, 1972) p. 73. Some other academic relatives of top political leaders are Aliev's brother Khasan, a full member of the Azerbaidzhan Academy of Sciences, and his recently deceased wife; Grishin's son who is apparently a graduate student at *IMEMO*; Chernenko's daughter, a specialist in the philosophical aspects of biology at the Institute of Marxism–Leninism; Kunaev's half-brother, president of the Kazakh Academy of Sciences, Solomentsev's son, who is rector of a machine tool institute in Moscow; Ustinov's defence physicist son, a corresponding member of the Academy and State Prize winner; and Fedorchuk's son who is a mathematician at Moscow State University. A. Rahr, 'Biographies of Politburo members', *Radio Liberty Supplement*, no. 1, 2 March 1984; *Kommunist*, 7/83, pp. 123–5; *Izvestiia*, 17 May 1985, p. 6, *Ekonomicheskaia gazeta*, no. 29 (July 1985) p. 6.

27. See above p. 25.

28. See Boris Vasil'evich Shcherbitsky's book, written with V. S. Tarasovich, *Planovoe upravlenie razvitiem nauki i tekhniki v soiuznoi respublike* ('Naukova dumka', Kiev, 1981).

29. *Kommunist*, 12/79, p. 35.

30. *Khrushchev Remembers. The last testament*, op. cit., p. 64.

31. Medvedev (1969) op. cit., pp. 93–5; Gustafson (1981) op. cit, p. 88.

32. In his memoirs Khrushchev seems to place the phone call in 1961, when the breaking of the moratorium was being discussed. One suspects that Sakharov's account is more accurate. A. D. Sakharov, *Sakharov Speaks*, Collins and Harvill, London, 1974, pp. 32–4; *Khrushchev Remembers. The last testament*, op. cit., pp. 68–71.

33. *Khrushchev Remembers. The last testament*, op. cit., pp. 69–70. Sakharov paraphrases Khrushchev's comment thus: 'Sakharov is a good scientist. But leave it to us, who are specialists in this tricky business, to make foreign policy . . . I would be a slob, and not chairman of the

Council of Ministers, if I listened to the likes of Sakharov.' Sakharov, op. cit., p. 33.

34. It should be mentioned that Kurchatov had earlier tried to work through the Academy hierarchy himself, but had given up and got Nikolai Semenov, at that time a more prestigious scientist, to write directly to the government. The beginning of the war ensured that Semenov got no response. D. Holloway, *Entering the Nuclear Arms Race. The Soviet decision to build the atomic bomb, 1939–45*, International Security Studies Program Working Paper No. 9, Wilson Center, Washington, DC, 1979, p. 17.

35. E. A. Beliaev, *KPSS i organizatsiia nauki v SSSR* (Politizdat, Moscow, 1982) p. 65; *Nauka i zhizn'*, 2/83, p. 17.

36. I. Golovin, *Academician Igor Kurchatov* ('Mir', Moscow, 1969) p. 56; *VAN*, 12/82, p. 88.

37. Gustafson, 1981, op. cit., p. 92.

38. E. Andronikashvili, *Nachinaiu s El'brusa (Tvorcheskie portrety uchenykh)* ('Metsniereba', Tbilisi, 1982) pp. 58–62. At a later stage Andronikashvili turned again to Kurchatov to get a nuclear reactor for his Tbilisi institute. Ibid, pp. 77–8.

39. Ibid., pp. 83–4.

40. Ibid., pp. 236–9, 288.

41. Jerry Hough provides indirect evidence that despite the minor role that Central Committee members play in formal Central Committee plenums, they do have a more significant informal role. He refers to an industrial worker who received regular Politburo reports in order to keep up-to-date with major policy issues and who participated in commissions in which decisions on the improvement of industrial administration were worked out. Indirect evidence further indicates that at formal sessions those few members who get the opportunity to speak use the forum in much the same way as Party Congresses are used, that is, to make pleas for special treatment for one's region or branch. Hough cites Brezhnev to the effect that these pleas are taken into consideration, and that a conscious effort is made to obtain the opinions of Central Committee members on policy issues. Hough and Fainsod, op. cit., pp. 461–5. For similar views, see R. J. Hill and P. Frank, *The Soviet Communist Party* (Allen and Unwin, London, 1981), pp. 64–5; V. E. McHale and J. P. Mastro, 'The Central Committee of the CPSU. Analysis of composition and long-term trends', paper presented to the 66th Annual Meeting of the American Political Science Association, Los Angeles, 8–12 September 1970.

42. Ther is nothing very striking about scientists' participation in formal plenums. I have looked at plenums since September 1965. Between then and 1971, Keldysh, president of the Academy of Sciences, spoke seven times, there having been twenty four plenums in that period. He did not speak again, and his successor Aleksandrov has spoken only three times (February 1976, on the occasion of the 24th Party Congress, the November 1979 planning and budget plenum, and the April 1984 plenum on educational reform). The other scientists to have spoken at plenums are Fedoseev (April 1968, June 1969, April 1985), Inozemtsev

(November 1971), Paton (May 1972 and November 1978), Marchuk (October 1976) and Ovchinnikov (May 1982). There is usually no obvious linkage between the speaker's post and the theme of the plenum.

43. *Spravochnik partiinogo rabotnika*, vyp. 21 (Politizdat, Moscow, 1981) p. 493.

44. The indicated reduction is in one way misleading, since some of those making the list in 1951 did so again in 1956, but were no longer in academic positions. Thus the 1951 full members were the philosopher Pospelov, the historian Pankratova and the economist Rumiantsev; the candidate members the philosophers G. F. Aleksandrov and Stepanova and the economist Ostrovit'ianov. In 1956 all the full members retained their membership, but Rumiantsev had gone to be chief editor of *Kommunist* and Pospelov was now a Central Committee secretary, neither academic posts. Aleksandrov and Stepanova both dropped out as candidates, although apparently remaining in academic work.

45. Iangel' is a mysterious character. Vladimirov claims that he was a German colleague of Werner von Braun who was captured by the Soviets towards the end of the war. Like so many of his compatriots he was put to work in a prison research centre. He went on to make his home and career in the Soviet Union. Vladimirov, op. cit., p. 44. This story is accepted by Zhores Medvedev. Medvedev (1979) op. cit., p. 99. James Oberg states that Iangel' was in fact born in Siberia, although of German stock. While he did work with von Braun during the war, he did so as a Soviet spy. J. E. Oberg, *Red Star in Orbit* (Harrap, London, 1981) p. 45. It certainly seems odd that the Soviets would go to the trouble of dreaming up a whole false biography for Iangel', including the purely Russian patronymic Kuz'mich and a classic *vydvyzhenets* career, or indeed that they would have loaded with honours even a POW who had done so much for the Soviet Union. The closest he gets to spying in his official biographies is a trip to the US in 1938 as the aviation expert on a trade delegation. *VAN*, 10/81, p. 117; V. Gubarev, *Konstruktor. Neskol'ko stranits iz zhizni Mikhaila Kuz'micha Iangelia* (Politizdat, Moscow, 1977).

46. *Deputaty Verkhovnogo Soveta SSSR. Deviaty sozyv* (Moscow, 1979), p. 270.

47. Some other famous names, such as Mitin and Iudin, have been left out of this survey, because the posts they held while in the Central Committee were not in research or science management.

48. G. M. Kornenko of the Ministry of Foreign Affairs; V. V. Zagliadin, O. B. Rakhmanin and A. S. Cherniaev of the Central Committee apparatus; and foreign affairs commentators A. E. Bovin and G. A. Zhukov. See E. Teague, 'The Central Committee and Central Auditing Commission elected at the 26th Congress of the CPSU', *Radio Liberty*, RL 171/81, 28 April 1981, pp. 12–13. Jerry Hough claims that the foreign affairs establishment has gained more in terms of Central Committee representation than any other group since the death of Stalin, but includes an enormously wide range of professions, including army officers and defence industry workers, in his 'foreign affairs' category. Hough and Fainsod, op. cit., p. 458.

49. It might be possible to argue that Semenov's election was more closely connected with his championing of the 'anti-applied research' wing of the Academy of Sciences in the great debate of the late 1950s, which culminated in the removal from the Academy of its technical institutes in 1961. However I am not convinced that the 1961 reorganisation was a victory for Semenov's faction. See Fortescue, *The Academy Reorganized*, op. cit., ch. 1.

50. See the series of articles by William Kucewicz in the *Wall Street Journal* in April–May 1984, especially 1 May 1984.

51. He did not speak at any Central Committee plenum after his retirement, although he remained active in Academy affairs.

52. According to the Academy Statute at the time, only a full academician could be Chief Scientific Secretary.

53. He appeared to be overshadowed by his predecessor, the veteran A. V. Topchiev. For example, at the November 1961 General Meeting, Topchiev spoke but Fedorov did not.

54. H. Kraus, *The Composition of Leading Organs of the CPSU (1952–1976)*, supplement to the Radio Liberty Research Bulletin, 1976, p. 37.

55. For perhaps the best summary of the structure and functions of the apparatus, see Hough and Fainsod, op. cit., pp. 409–48.

56. Previously, in the 1920s, a deputy head of the Agitprop Department was apparently responsible for the natural sciences. Y. M. Rabkin, 'On the origins of political control over the content of science in the Soviet Union' (an interview with A. Kol'man), *Canadian Slavonic Papers*, vol. 21, no. 2, 1979, p. 228; Kol'man, op. cit., p. 161; M. Fainsod, *How Russia is Ruled* (Harvard UP, Cambridge Mass., 1967) revised edition, pp. 193–4; L. V. Ivanova, *Formirovanie sovetskoi nauchnoi intelligentsii (1917–27gg.)* ('Nauka', Moscow, 1980) p. 52. In 1930 Agitprop was split into Culture and Propaganda and Agitation and Mass Campaigns. One of the three sectors in Culture and Propaganda was the Science and Education sector. Beliaev, op. cit., p. 41.

57. *Partiinoe stroitel'stvo*, 11/35, p. 47. See also *Vestnik Kommunisticheskoi Akademii*, 7/35, p. 3; Beliaev, op. cit., p. 45; V. K. Beliakov and N. A. Zolotarev, *Organizatsiia udesiateriaet silu. Razvitie organizatsionnoi struktury KPSS. 1917–1974gg.* (Politizdat, Moscow, 1975) p. 73; Kol'man, op. cit., p. 73; Fainsod, op. cit., p. 194.

58. Beliaev, op. cit., p. 46. A Science Department apparently remained in existence, but within the new Cadres Administration. Hahn, op. cit., p. 199; *Partiinoe stroitel'stvo*, 13/39, p. 6.

59. Beliaev, op. cit., p. 59; Hahn, op. cit., p. 201; *Istoriia Kommunisticheskoi Partii Sovetskogo Soiuza*, vol. 5, bk 1 (Politizdat, Moscow, 1970) p. 448. *Istoriia KPSS* makes no mention of the war-time subordination of the department. Beliaev cites *Istoriia* for its establishment, but gives no source for his statement that it was within the Agitation and Propaganda Administration. However by early 1948 it was being so described in the Soviet press. *Pravda*, 22 April 1948, p. 3.

60. Hahn, op. cit., pp. 205–6; *Istoriia KPSS*, vol. 5, bk 2 (Politizdat, Moscow, 1980) pp. 221, 249.

61. Hahn, op. cit., p. 207.

62. Fainsod, op. cit., p. 202.
63. Hough and Fainsod, op. cit., p. 416.
64. Zimianin, a Central Committee secretary for ideology, attended a 1976 Academy General Meeting, the 1980 reelection of Aleksandrov as president and the 1983, 1984 and 1985 Annual General Meetings; Ponomarev the 1982 AGM; and Suslov the retirement of Keldysh and the election of Aleksandrov as president.
65. *Politichesky dnevnik 1964–70*, op. cit., pp. 119–21.
66. Ibid., p. 663.
67. Zaslavsky, op. cit., p. 165. Zaslavsky quotes him as having said: 'Soviet history does not contain, and never has contained negative pages.'
68. Katsenelinboigen (1980) op. cit., pp. 22, 158; Zemtsov, op. cit., p. 57; Nekrich, op. cit., p. 291; R. Medvedev, op. cit., p. 217.
69. *New York Times*, 22 July 1968, p. 15.
70. Nekrich, op. cit., p. 291.
71. *Pravda*, 14 March 1984, p. 3.
72. A. Rahr, 'Brezhnev's former party theorist Trapeznikov retires on pension', *Radio Liberty*, RL 350/83, 19 September 1983.
73. *VAN*, 12/83, p. 13.
74. 'Znanie', Moscow, 1965. For some other publications, see *Partiinaia zhizn'*, 6/80, pp. 17–24; *Oktiabr'*, 8/77, pp. 165–79; *Sovetskaia politicheskaia sistema v usloviiakh razvitogo sotsializma* ('Mysl'', Moscow, 1975).
75. *Voprosy istorii KPSS*, 11/83, p. 158.
76. For example, *Politicheskaia ucheba v nauchnom kollektive* (Politizdat, Moscow, 1974); *Sovetskaia Rossiia*, 28 August 1970, p. 2.
77. For example, a 1977 General Meeting of the Ukrainian Academy of Sciences and a 1982 meeting of the Ukrainian party-economic *aktiv* on scientific and technical progress, as well as the 1977 meeting of the presidents of the socialist Academies addressed by Brezhnev.
78. According to Hough and Fainsod the Ministry of Medical Industry is supervised by the Science Department. Hough and Fainsod, op. cit., p. 416.
79. Economics, health, *vuzy*, philosophy, specialist secondary educational institutions, and the social sciences.
80. According to Hough and Fainsod the Academy for Construction and Architecture and the Union of Architects come under the Central Committee's Construction Department. Ibid., p. 413.
81. S. Reznik, *Doroga na eshafot* ('Tret'ia volna', Paris and New York, 1983) p. 18.
82. Berg, op. cit., p. 249. Note that Jerry Hough specifically states that *konsul'tanty*, despite the name, are full-time staff. Hough, 'Soviet policy-making', op. cit., p. 175. It is also worth noting that D. K. Simes stresses that academics are used as 'consultants' strictly in the English sense of the word, not as policy makers. D. K. Simes, 'National security under Andropov', *Problems of Communism*, vol. 32, no. 1, January–February 1983, p. 37.
83. There is a hint of the transfer of personnel between the Science Department and GKNT. L. I. Mazur worked in the Science Department in 1967, and appears as a GKNT staff member in 1971. In neither case is his area of work mentioned.

84. G. Hodnett and V. Ogareff, *Leaders of the Soviet Republics, 1955–1972*, and V. Ogareff, *Leaders of the Soviet Republics, 1971–1980*, Department of Political Science, RSSS, Australian National University, Canberra, 1973 and 1980.
85. See below, pp. 118–19.
86. Beliaev, op. cit., p. 75.
87. For example, *Moskovskaia pravda*, 25 October 1969, pp. 1–2.
88. *Kommunist Sovetskoi Latvii*, 5/82, p. 9. It should be admitted that a recent survey of Soviet emigres showed quite a high percentage of branch researchers giving an important role to the Science Department. R. F. Miller, 'The role of the Communist Party in Soviet research and development', *Soviet Studies*, vol. 37, no. 1, January 1985, p. 48. It does seem possible however that the respondents were wrong, and in fact commenting on the importance of the Central Committee apparatus in general. They were not asked the question with regard to any other department. In the literature and conversation the Central Committee is usually referred to in very vague terms, with no differentiation between departments. Indeed it is used as virtually a generic name for 'higher authorities' or 'party leadership'. Note that Mark Popovsky seems not to have known even what department it was he was called to after the publiction of one of his essays on Nikolai Vavilov. M. Popovsky, *Delo akademika Vavilova* ('Ermitazh', Ann Arbor, 1983) p. 22.
89. Miller (1985) op. cit., p. 33. Note also that Hough and Fainsod report one former laboratory head of a metallurgical institute among Central Committee instructors they have identified. Hough and Fainsod, op. cit., pp. 428–9.
90. According to Voslensky, two typed pages are prepared for the discussion of resolutions, while members receive background papers five pages in length. Cited in A. von Borcke, 'Dilemmas of single party rule in the Soviet Union', *Problems of Communism*, vol. 31, no. 3, May–June 1982, p. 61.
91. In the case of Central Committee plenums it is openly admitted that resolutions are previously examined by the Politburo. *Kommunist*, 12/79, p. 34. Although they gloss over the Politburo stage, Kunaev and Shevardnadze, the first secretaries of the Kazakh and Georgian parties, give reasonable detail on the preparation of plenums in *Partiinaia zhizn'*, 16/81, p. 11; *Zaria Vostoka*, 15 May 1982, p. 1. For more details, see G. Hodnett, 'The pattern of leadership politics', in S. Bialer (ed.), *The Domestic Context of Soviet Foreign Policy* (Westview, Boulder, 1981) pp. 99–101; O. Kh. Moiseishin, *Deistvennost' partiinogo kontrolia. Iz opyta Kompartii Belorussii* ('Belarus', Minsk, 1982) p. 12. For details on the preparation of the 1958 draft Central Committee resolution on the journal *Voprosy istorii*, see Hough in Skilling and Griffiths, op. cit., p. 83.
92. D. K. Simes makes the interesting point that, in foreign policy anyway, journalists are far more influential with the party hierarchy than academics. This is because journalists, with their greater access to the public, must meet more rigorous security clearance procedures and are generally considered more reliable. Even outside foreign policy this

was true in past years – what influential social scientists there were, the likes of Pospelov, Mitin or Rumiantsev, had considerable journalistic experience. It seems less true now, although such editors as Afanas'ev of *Pravda* and Kosolapov of *Kommunist* have academic pretensions. Simes, op. cit., p. 37.

93. Z. Mlynar, *Night Frost in Prague* (Karz, New York, 1980) pp. 56–7.
94. Ibid., pp. 57, 136.
95. A. Katsenelinboigen, 'Planning and Soviet science', *Survey*, vol. 23, no. 1 (102) (Winter 1977–8) p. 43; Azbel, op. cit., p. 251; Lubrano in Remnek, op. cit., p. 64; *Ideino-politicheskoe vospitanie*, op. cit., p. 158; *VAN*, 12/83, p. 4; A. Brown, 'Political science in the Soviet Union: a new stage of development?', *Soviet Studies*, vol. 36, no. 3 (July 1984) pp. 340–1; Hough in Skilling and Griffiths, op. cit., p. 83.
96. Belitsky, op. cit., p. 38. See also Gallagher and Spielmann, op. cit., p. 53.
97. Zemtsov, op. cit., p. 33.
98. *Izvestiia*, 18 February 1984, p. 2.
99. Pilipenko was a party worker from Iaroslavl' who first made an appearance in Moscow in the mid-1960s as head of the Universities Administration of the RSFSR Ministry of Higher Education. He is a doctor of philosophical sciences, and regularly publishes articles and monographs on ideological subjects with a scientific twist, for example, *Dialektika neobkhodimosti i sluchainosti* ('Mysl'', Moscow, 1981).
100. See above, pp. 28–29.
101. Belitsky, op. cit., pp. 42–3.
102. The conservatives tend to base their opposition on calls for greater mass participation and criticism of the 'liberals', overly 'technocratic' and professional elitist views. See J. F. Hough, 'Soviet succession: issues and personalities', *Problems of Communism*, vol. 31, no. 5, September–October 1982, p. 22; A. Brown, 'Pluralism, power and the Soviet political system: a comparative perspective', in S. G. Solomon, op. cit., pp. 61–107, esp. pp. 82–91. See also A. Brown, 'Andropov: discipline *and* reform?', *Problems of Communism*, vol. 32, no. 1 (January–February 1983) pp. 18–31.
103. Hough, 'Soviet policy-making', op. cit., p. 175.
104. This contrasts with the diplomatic corps of the Ministry of Foreign Affairs which, with the rotation of its best officers abroad, does not have a regular top-quality policy group resident in Moscow. Ibid., pp. 178–9.
105. A. Pravdin (interviewed by M. Matthews), 'Inside the CPSU Central Committee', *Survey*, vol. 20, no. 4 (93) (Autumn 1974) pp. 97–8.
106. Brown (1984) op. cit., p. 326.
107. There is some dispute in the literature over whether it is always necessary to work through the Central Committee apparatus to get access to the Politburo. Karen Dawisha reports a Ministry of Foreign Affairs official as claiming that the ministry had to send proposals for Politburo discussion through the Central Committee apparatus even after Gromyko had become a Politburo member. Archie Brown, citing Mly-

nar's experiences in Czechoslovakia, suggests that a minister has to work through the apparatus *unless* he is a Politburo member, and Shtromas claims that it is only when the military has no representatives on the Politburo that it relies on the Central Committee. Birman reports a case of the head of GKNT giving documents directly to Politburo members. K. Dawisha, 'The limits of the bureaucratic politics model: observations on the Soviet case', *Studies in Comparative Communism*, vol. 13, no. 4 (Winter 1980) p. 322; Brown in Solomon, op. cit., p. 88; Shtromas, op. cit., p. 116; Birman, op. cit., p. 43.

108. P. H. Solomon (1978) op. cit., p. 111. We do have one historical case that seems to confirm this finding. Proposals for the organisation of the Jet Propulsion Research Institute (*Reaktivny nauchno-issledovatel'sky institut*) were sent to the Central Committee for consideration and decision, but only after they had been gone over thoroughly by the scientists and military officers involved. We have the usual problem in this case that in the available reports we are given no hint as to what the 'Central Committee' actually means. *Voprosy istorii estestvoznaniia i tekhniki*, 2/84, pp. 112–13.

109. M. Voslensky, *Nomenklatura. Anatomy of the Soviet ruling class* (Bodley Head, London, 1984).

110. *Times Higher Educational Supplement*, 26 March 1982, pp. 12–13.

111. Zemtsov, op. cit., p. 40; Berg op. cit., p. 289; Kneen (1984) op. cit., p. 70.

112. When certain natural science disciplines were particularly sensitive this would have applied to them also. We know that specific posts are added to or moved to higher-level *nomenklatury* at times of particular sensitivity. See below, pp. 148, 207.

113. *Nauchno-organizatsionnaia deiatel'nost' akademika A. F. Ioffe. Sbornik dokumentov* ('Nauka', Leningrad, 1980) p. 127. Tikhomirov was chairman of the reported meeting, but no details on who he was are available.

114. Azbel, op. cit., pp. 316–21.

115. Nekrich, op. cit., p. 347.

116. Zemtsov, op. cit., p. 58.

117. For a case of a primary party organisation secretary refusing to sign such a reference, see Medvedev (1972) op. cit., p. 103.

118. Mark Popovsky puts the role of these old pensioners in more sinister context. Popovsky (1978) op. cit., p. 152.

119. Medvedev (1972) op. cit., pp. 103, 249. See also Popovsky (1978) op. cit., pp. 155–6. For a case of the Science Department overriding a decision not to grant permission for travel, see Berg, op. cit., p. 184. See the same source for a case of the Central Committee apparatus refusing to grant permission to travel. Ibid., p. 248.

120. For some accounts of such problems, see Medvedev (1972) op. cit., p. 141; *Science*, 3 October 1975, pp. 34–5. The same things are said of academic exchanges. *New York Times*, 17 May 1983, pp. C1, C6.

121. *Kommunist*, 18/68, p. 40. See also *Ideino-politicheskoe vospitanie*, op. cit., pp. 68, 101.

122. It should also be mentioned that the Soviets are happy to put pressure

on conference organizers to remove even non-Soviet delegations and papers that are not to their liking. *VAN*, 2/58, p. 76; *Sovetskoe gosudarstvo i pravo*, 6/66, p. 135.

123. Z. A. Medvedev, *Nuclear Disaster in the Urals* (Vintage, New York, 1980) p. 23; *Literaturnaia gazeta*, 4 January 1978, p. 11.
124. Azbel, op. cit., p. 281.
125. Medvedev (1972) op. cit., p. 484.
126. Nekrich (op. cit); Reznik, op. cit., pp. 5–25.
127. Jones, op. cit., p. 177.
128. *VAN*, 4/70, p. 56.
129. *VAN*, 9/80, p. 35.
130. A similar example is an article written by the Lithuanian Central Committee secretary for industry, about new production facilities in Academy institutes and their involvement in R&D programmes. *Ekonomicheskaia gazeta*, no. 17 (April 1980) p. 9.
131. The Propaganda Department is responsible for philosophical and methodological seminars. *Marksistsko–leninskoe obrazovanie i nauchnoe tvorchestvo*, op. cit., p. 157; *Kommunist*, 16/79, p. 42; *Voprosy filosofii*, 6/75, p. 161. The last reference is particularly striking, in that it suggests that seminars in Moscow are run by the Central Committee Propaganda Department and the *gorkom* Science Department.
132. The Science Department is said to have had 41 to 43 'responsible workers', plus 35 other personnel in the early 1970s. Pravdin, op. cit., p. 95. In 1979 the CIA listed 35 members of the Science Department. *Directory of Soviet Officials. National Organizations. A reference aid* (Directorate of Intelligence, Washington DC, 1979) pp. 31–2.
133 R. C. Wood, 'Scientists and politics: the rise of an apolitical elite', in R. G. Gilpin and C. Wright (eds), *Scientists and National Policy Making* (Columbia UP, NY and London, 1964).
134. See above pp. 48–49.
135. *Pravda*, 25 February 1976, p. 6.
136. *VAN*, 10/76, p. 13; 9/76, p. 11. See also academicians Emanuel' and B. S. Sokolov, *VAN*, 5/76, p. 47; 7/76, p. 85.
137. *VAN*, 6/77, pp. 8–9, 69–73. The alleged objectivity of the Academy, derived from its lack of branch subordination, is an argument often put forward by those pushing for greater powers for it. See, for example, *VAN*, 9/75, p. 14; 5/80, pp. 102–4. For more details on the expanded role of the Academy, see Fortescue, *The Academy Reorganized*, op. cit., pp. 50–7; 'Akademiia nauk – gotovnost' k kompromissu', *Obozrenie*, no. 10 (July 1984) pp. 7–10.
138. *VAN*, 7/81, p. 6.
139. *VAN*, 6/78, p. 59.
140. *Izvestiia*, 4 July 1984, p. 2.
141. J. R. Thomas and U. M. Kruse-Vaucienne, 'Soviet science and technology: an introduction', *Survey*, vol. 23, no. 1 (102) (Winter 1977–8) p. 4; L. R Graham, 'The role of the Academy of Sciences', ibid., p. 130; Miller (1985) op. cit., p. 53.
142. For details, see Fortescue, *The Academy Reorganized*, op. cit., pp. 33–5.

143. The best example being Brezhnev's tirade at the November 1979 Central Committee plenum.
144. The new system of regional planning and management of science, which is perhaps the most striking example of Academy–party cooperation, is discussed in the next chapter.
145. *VAN*, 6/83, p. 41; 7/83, p. 44.
146. Such planning is supposedly based on long-term forecasts, which are worked out primarily by scientists using *ekspertiz* methods, under the supervision of a special joint scientific council of GKNT and the Academy, headed by Academy vice-president Kotel'nikov. The resulting long-term programme is divided into five-year plans for fundamental research, the responsibility of the Academy, applied research under GKNT, and *vnedrenie* under Gosplan. If the system were working properly the forecasting and fundamental and applied research plans, run by the Academy and GKNT, should be providing the lead for the *vnedrenie* and production plans of Gosplan. For details on programme planning, see S. Fortescue, 'Project planning in Soviet R & D', *Research Policy* vol. 14 (1985), pp. 267–82; Amann and Cooper, op. cit., pp. 477–91.
147. For details, see Fortescue, *The Academy Reorganized*, op. cit., ch. 4.
148. Shcherbakov, op. cit., p. 30.
149. V. P. Rassokhin is the champion of the 'fourth link' proposal. See his articles in V. A. Trapeznikov (ed.), *Upravlenie razvitiem nauki i tekhniki* ('Ekonomika', Moscow, 1980) pp. 58–68; *VAN*, 11/79, pp. 43–53; 11/81, pp. 53–61; *Izvestiia*, 25 May 1971, p. 3; *EKO*, 1/80; 2/81. For references to other proposals, see Graham, 1977–8, op. cit., p. 120; *Kommunist*, 4/83, p. 71; I. A. Glebov and I. I. Sigov, *Sovershenstvovanie upravleniia fundamental'nymi issledovaniiami v krupnom gorode* ('Nauka', Leningrad, 1983) p. 153; Pokrovsky, op. cit., pp. 177–8. For what seems to be a case of Marchuk, chairman of GKNT, publicly joining the campaign for significantly greater powers for GKNT, see *Partiinaia zhizn'*, 1/85, p. 33.
150. Although Soviet commentators are well aware of this aspect of the problem. Pokrovsky, op. cit., p. 126.
151. Fortescue (1986) op. cit., p. 274.
152. See Rigby, Brown and Reddaway, op. cit; Breslauer, op. cit.
153. For an interesting account of the 'collegial' approach to administration and its negative effect on decisive policy-making, see Jones, op. cit.
154. See the demands made of the Academy's Urals Science Centre for the coordination of all R&D organisations in the region regardless of administrative subordination in the Central Committee decree of October 1983. The Urals Science Centre has no statutory powers or financial or administrative resources to enforce such coordination. *Pravda*, 20 October 1983, p. 1; *VAN*, 3/84, p. 60.
155. For a good example, see the August 1983 Central Committee and Council of Ministers decree on scientific-technical progress. *Pravda*, 28 August 1983, p. 1.

CHAPTER 3

1. In this chapter I am concerned, in a rather undifferentiated way, with all levels of the party apparatus between the Central Committee and primary party organisation, essentially the *raikom, gorkom* and *obkom*. I will use both the terms 'regional' and 'local', but without wanting in doing so to make the usual distinction between *obkom*-level party organs and the lower level *raikomy*. To the extent that I have spoken of republican Central Committees at all, they were included in the last chapter.

2. That particular campaign seems designed more to shift the focus of centralised decision-making from the branch-based state apparatus to the equally centralised but regionally-organised party apparatus.

3. A. G. Frank, 'Goal ambiguity and conflicting standards: an approach to the study of organization', *Human Organization*, vol. 17, no. 4 (Winter 1958–9) pp. 8–13; Rigby (1976) op. cit., pp. 38–40.

4. J. F. Hough, *The Soviet Prefects. The local party organs in industrial decision-making* (Harvard UP, Cambridge, Mass., 1969).

5. L. R. Graham, *The Soviet Academy of Sciences and the Communist Party. 1927–1932* (Princeton UP, 1967); A. V. Kol'tsov, *Razvitie Akademii nauk kak vysshego nauchnogo uchrezhdeniia SSSR. 1926–1932* ('Nauka', Leningrad, 1982), pp. 20, 63–70, 186, 194.

6. S. Fitzpatrick, *Education and Social Mobility in the Soviet Union* (Cambridge UP, 1979) pp. 81, 90; I. P. Barmin, *Iz opyta raboty KPSS i sovetskogo gosudarstva po sozdaniiu kadrov sovetskoi intelligentsii (1928–1933)* (Izd-vo Moskovskogo un-ta, Moscow, 1965); *Voprosy istorii KPSS*, 2/66, pp. 29–42; Kol'tsov, op. cit., pp. 168–170.

7. V. I. Salov, 'Iz istorii Akademii nauk SSSR v pervye gody Velikoi Otechestvennoi voiny (1941–1943)', *Istoricheskie zapiski* (Izd-vo AN SSSR, Moscow, 1957) pp. 3–20; G. L. Sobolev, *Uchenye Leningrada v gody Velikoi Otechestvennoi voiny, 1941–1945* ('Nauka', Moscow and Leningrad, 1966); *Voprosy istorii KPSS*, 9/76, pp. 116–25.

8. B. I. Kozlov (ed.), *Organizatsiia i razvitie otraslevykh nauchno-issledovatel'skikh institutov Leningrada. 1917–1977* ('Nauka', Leningrad, 1979) p. 90; Smolkina, op. cit., pp. 34–36.

9. Barrington Moore, op. cit., p. 135.

10. *Kommunist*, 18/54, pp. 53–76.

11. *KPSS v rezoliutsiiakh i resheniiakh s"ezdov konferentsii i plenumov TsK*, pt 4, 1954–60 (Moscow, 1960) 7th edition, pp. 91–109.

12. *Kommunist Ukrainy*, 8/55, pp. 46–55; *Partiinaia zhizn'*, 12/57, pp. 21–7; 10/58, pp. 25–35; 11/58, pp. 8–15.

13. Medvedev (1972), op. cit., pp. 499–500, 567–8. For criticism of the Obninsk *gorkom*, see *Kommunist*, 18/68, p. 45.

14. *Sovetskaia Rossiia*, 28 August 1970, p. 2.

15. *Politichesky dnevnik II*, op. cit., p. 367; Ianevich, op. cit., pp. 149–51.

16. Although regional party bodies will usually still be involved in political purges or campaigns within research institutes. For example, the involvement of the Moscow *gorkom* in the 1976 purge of *TsEMI* and the meeting of the buro of the Novocheremushki *raikom* to deal with

Soviet sociology in 1969. Zemtsov, op. cit., p. 50; Katsenelinboigen (1980) op. cit., p. 130.

17. Kanygin and Botvin claim that regional party organs have been held particularly responsible for science management since the 24th Party Congress in 1971. Iu. M. Kanygin and V. A. Botvin, *Problemy razvitiia i ispol'zovaniia nauchnogo potentsiala krupnykh gorodov* ('Naukova dumka', Kiev, 1980) p. 172.
18. R. A. Belousov and G. B. Khromushin (eds), *Partiinoe rukovodstvo ekonomikoi* ('Ekonomika', Moscow, 1981) p. 106; Ianovsky (1979) op. cit., p. 198. For a list of Novosibirsk *obkom* and Sovetsky *raikom* decisions on scientific matters, see ibid., pp. 274–7.
19. M. F. Tolpygo, *Partiinye komitety i tekhnichesky progress* ('Moskovsky rabochy', Moscow, 1975) p. 17.
20. *Pravda*, 25 May 1966, pp. 2–3.
21. *Zaria Vostoka*, 5 June 1982, p. 6.
22. *Partiinaia zhizn'*, 19/81, p. 23.
23. *Zaria Vostoka*, 5 June 1982, p. 6.
24. *Ekonomicheskaia gazeta*, no. 30 (July 1969) p. 6.
25. *Kommunist*, 6/78, pp. 77–78; *Sotsialisticheskaia industriia*, 27 December 1977, p. 2.
26. *Pravda*, 22 September 1974, p. 2. It is unusual for a party organ to use the word *obiazat'*. It is generally considered that the party's formal powers are purely recommendatory.
27. *Ekonomicheskaia gazeta*, no. 34 (August 1973) p. 10.
28. A. I. Sliusarenko, *KPSS i progress nauki v razvitom sotsialisticheskom obshchestve*, ('Vishcha shkola' Kiev, 1978) p. 64.
29. Ianovsky (1979) op. cit., pp. 274–5.
30. *Pravda*, 25 May 1966, pp. 2–3.
31. Kanygin and Botvin, op. cit., p. 152.
32. *Sovetskaia Rossiia*, 4 September 1966, p. 2; *Ekonomicheskaia gazeta*, no. 38 (September 1966) pp. 6–7; Kozlov, op. cit., pp. 126–8.
33. *Kommunist*, 9/80, p. 47.
34. *Pravda*, 27 December 1968, p. 2; 6 September 1973, p. 2.
35. *Kommunist*, 9/80, p. 39.
36. S. V. Sharenkov, *Raikom i nauchno-tekhnicheskaia intelligentsiia* (Lenizdat, Leningrad, 1981) pp. 44–5.
37. *Kommunist*, 9/80, p. 43. See also *VAN*, 8/82, pp. 34–44.
38. *Partiinaia zhizn'*, 2/80, pp. 6–7; 9/82, p. 56; *Trud*, 9 August 1974, p. 2.
39. *Times Higher Educational Supplement*, 26 March 1982, pp. 12–13; Medvedev (1979) op. cit., p. 124; Zemtsov, op. cit., p. 40.
40. Sharenkov, op. cit., p. 8. It is worth noting Peter Kneen's observation that *nomenklatury* are flexible and posts can be added to a higher party organ's *nomenklatura* at times of particular priorities or difficulties. Kneen, op. cit., pp. 70–1. For an example, see *Ideino-politicheskoe vospitanie*, op. cit., p. 66.
41. *Kommunist*, 9/80, p. 42.
42. *Pravda Ukrainy*, 6 February 1972, p. 2.
43. *Partiinaia zhizn'*, 9/77, p. 35; 21/78, p. 41; *VAN*, 11/79, p. 52; 5/81, pp. 62–7. The Proletarsky *raikom* in Moscow 'orients' every production

collective to have 'contracts of cooperation' with a scientific institution. *Partiinaia zhizn'*, 1/77, p. 39.
44. *Pravda*, 25 October 1982, p. 2.
45. *Kommunist*, 9/81, pp. 48–58.
46. Note that local party authorities were inclined to support the innovatory organization *Fakel* over the opposition of bureaucratic agencies. See Lowenhardt (1974) op. cit.
47. Not all Western commentators agree that regionalisation increases the role of the regional party apparatus. See J. F. Hardt and G. D. Holliday, 'Technology transfer and change in the Soviet economic system', in Fleron, op. cit., p. 217.
48. *VAN*, 12/80, pp. 17–27.
49. *VAN*, 8/81, pp. 35–62.
50. For more recent articles on the L'vov system, see *VAN*, 10/83, pp. 92–100; *Kommunist*, 7/83, pp. 54–63.
51. *VAN*, 9/81, pp. 38–46; 6/83, p. 4; *Leningradskaia pravda*, 31 March 1983, p. 1; 29 May 1983, p. 2.
52. *VAN*, 9/81, p. 44.
53. *VAN*, 12/81, pp. 38–9.
54. *Kommunist*, 2/81, pp. 31–2.
55. *VAN*, 12/80, p. 23.
56. *VAN*, 9/81, p. 44.
57. Belousov and Khromushin op. cit., p. 73.
58. Although some writers criticise the overly close links with narrow interests of the various branches of industry that GKNT's organizational structure produces, there are signs of cooperation between the Academy and GKNT for purposes of overcoming 'departmental' barriers to progress. See above p. 93.
59. *VAN*, 8/81, p. 40; 10/83, p. 93. The instrument-making complex of the L'vov system, which is one of four complexes and includes five scientific-production associations, had 890 people involved in its work in 1981 and had a budget of 2.5 million rubles. *Partiinaia zhizn*, 8/81, p. 51.
60. See the call for science centres in Kazakhstan. *VAN*, 7/82, p. 16.
61. Plus its Siberian branch, nine affiliates in Autonomous Republics, and the North Caucuses Science Centre of the All-Union and RSFSR Ministries of Higher Education.
62. See the defensiveness of Paton and Kotel'nikov in stressing that they do not see the programmes as the panacea for all problems, and Kotel'nikov's cryptic statement that not all those who had wanted to attend the 1981 seminar had been able to, 'for various reasons'. *VAN*, 8/81, p. 36; *Komsomol'skaia pravda*, 22 March 1980, p. 3.
63. *Pravda Ukrainy*, 30 November 1982, p. 2.
64. Ruble, op. cit., p. 39.
65. Glebov and Sigov, op. cit., p. 153; Pokrovsky, op. cit., pp. 181–2.
66. *Metodologicheskie problemy kompleksnykh issledovanii* ('Nauka', Novosibirsk, 1983) p. 248.
67. *Izvestiia*, 27 January 1985, p. 2; *Pravda* 26 July 1978, p. 2; Sigov, op. cit., p. 76.

68. *Sovetskaia Rossiia*, 11 March 1983, p. 1.
69. See below, p. 114.
70. Belousov and Khromushin, op. cit., pp. 64, 73.
71. *VAN*, 4/78, pp. 134–5; 10/78, pp. 59–64; 9/81, p. 140. See also *VAN*, 6/83, p. 38.
72. *VAN*, 7/78, pp. 7–8; 6/82, p. 72; 8/83, p. 19.
73. P. Taylor, 'Personnel changes in Leningrad oblast', *Radio Liberty*, RL 287/83, 29 July 1983. See also D. P. Hammer, 'Brezhnev and the Communist Party', *Soviet Union/Union Soviétique*, vol. 2, pt 1, 1975, pp. 9–11.
74. *Pravda*, 25 May 1966, p. 2–3.
75. *Pravda Ukrainy*, 6 February 1972, p. 2.
76. *Ekonomicheskaia gazeta*, no. 34 (August 1973) p. 10.
77. *Pravda*, 22 September 1974, p. 2. The delays clearly continue, judging by a 1984 article by the head of the Centre, Academician Vonsovsky. He is hopeful – although he does not sound very optimistic – that the 1983 decree on the Urals Science Centre will improve matters. *Izvestiia*, 18 February 1984, p. 2.
78. *Sovetskaia Rossiia*, 11 March 1983, p. 1.
79. Miller (1985) op. cit., pp. 47–9.
80. For a good introduction, see Hough and Fainsod, op. cit., ch. 13.
81. *Raionny komitet partii* (Politizdat, Moscow, 1974) 2nd edition, p. 238.
82. *Kommunist*, 11/72, p. 51–64.
83. *Voprosy filosofii*, 6/71, p. 141.
84. *VAN*, 4/75, p. 7; 7/78, pp. 7–8.
85. *Pravda*, 22 September 1974, p. 2.
86. Kozlov, op. cit., p. 153; *Ekonomicheskaia gazeta*, no. 34 (August 1973) p. 10.
87. In 1974 over one third of the *raion's* communists worked in research institutes and almost half its working population worked in science. *Kommunist*, 7/74, p. 43; V. I. Duzhenkov, *Problemy organizatsii nauki (regional'nye aspekty)* ('Nauka', Moscow, 1978) p. 45.
88. For accounts of the Institute of History episode, see Nekrich, op. cit., pp. 250–1. Radio Liberty claims that Chaplin was removed as first secretary of the Cheremushki *raikom* and sent as Ambassador to Vietnam following the infamous bulldozing of the open-air exhibition of unofficial artists, the same incident which supposedly ruined the career of Boris Iagodkin. *Radio Svoboda*, RS 221/83, 7 November 1983, p. 3.
89. See his article on science in *Kommunist*, 9/80, pp. 38–49.
90. See his article on science in *Kommunist*, 6/78, pp. 71–82.
91. For some publications, see A. P. Dumachev (with A. P. Solov'ev), *Za effektivnost' nauki* (Lenizdat, Leningrad, 1968); *Khozraschetnye ob"edineniia v promyshlennosti* (Lenizdat, Leningrad, 1972); *Partiinye organizatsii i proizvodstvennye ob"edineniia* (Politizdat, Moscow, 1976).
92. P. D. Stewart, *Political Power in the Soviet Union. A study of decision-making in Stalingrad* (Bobbs-Merrill, Indianapolis and New York, 1968) p. 181.
93. For example, Ianovsky (1974) op. cit.; M. P. Chemodanov, *Pre-*

vrashchenie nauki v neposredstvennuiu proizvodstvennuiu silu v prot-sesse stroitel'stva kommunizma ('Mysl", Moscow, 1966).

94. *Primeniaia individual'ny podkhod* (Politizdat, Moscow, 1982).
95. There are exceptions to the rule. Thus I. M. Cherepanov, once head of the Science Department of the Moscow *obkom*, became deputy chairman of the Moscow *oblispolkom* and chairman of the Moscow *oblast'* planning commission. V. Ia. Khodyrev, head of the Science Department of the Leningrad *obkom* in 1977, went on to become second secretary of the Leningrad *gorkom* and then *obkom*. See also Polunin above.
96. *Trud*, 5 February 1967, p. 3; *Pravda*, 22 September 1974, p. 2.
97. *VAN*, 1/77, p. 42; 8/81, p. 8.
98. *Ekonomicheskaia gazeta*, no. 51 (December 1966) p. 6. Interestingly Solov'ev later became head of the Labour Administration of both the Leningrad city and *oblast'* executive committees. It seems probable that he is a specialist in labour organization.
99. *Sovetskaia Rossiia*, 27 April 1971, p. 2.
100. *Sovetskaia Rossiia*, 4 October 1969, p. 2; *Sotsialisticheskaia industriia*, 18 April 1974, p. 2.
101. *Pravda*, 1 September 1983, p. 2.
102. *VAN*, 12/80, pp. 23–24; S. M. Pisarenko, *Tselevoe planirovanie i territorial'nye kompleksy* ('Naukova dumka', Kiev, 1983) p. 163.
103 *VAN*, 9/81, p. 44.
104. Sharenkov, op. cit., p. 76. Again there are exceptions. See the 1977 article by the head of the Science Department of the Voroshilovgrad *obkom*, a candidate of economic sciences, on branch research (although the focus of the article is on overcoming 'sectional interests' by setting up a special council in the Donetsk Science Centre of the Ukrainian Academy). *Pravda Ukrainy*, 12 November 1977, p. 2. His successor has written in the same vein. *Izvestiia*, 9 August 1979, p. 2. The head of the Science Department of the Tomsk *obkom* is deputy chairman of the *obkom's* council for the coordination of scientific research work, although heads of other departments are involved in some of the specific programmes managed by the council. *VAN*, 11/83, p. 51. This perhaps constitutes evidence that Science Departments are using the increased involvement of the Academy of Sciences in the regional management of science to increase their own powers.
105. See above, pp. 102–3.
106. *Partiinaia zhizn'*, 16/78, p. 21; 24/80, p. 34.
107. Hough (1969) op. cit., pp. 326–30.
108. *Partiinaia zhizn'*, 14/76, p. 44.
109. *Kommunist*, 9/80, p. 38.
110. Hough and Fainsod, op. cit., p. 502.
111. *Moskovskaia pravda*, 20 June 1965, p. 2.
112. *Pravda Ukrainy*, 26 January 1972, p. 2.
113. *Partiinaia zhizn'*, 8/77, pp. 39–41. See also *Partiinaia zhizn'*, 19/81, p. 23; 21/81, p. 50; Sharenkov, op. cit., pp. 25–36; *Izvestiia*, 27 September 1969, pp. 1, 3; 17 May 1972, p. 3.
114. *Pravda*, 19 September 1971, p. 3.

115. Kanygin and Botvin, op. cit., pp. 172–4.
116. *Ekonomicheskaia gazeta*, no. 20 (May 1977) p. 5; *Partiinaia zhizn'*, 10/78, pp. 42–3; I. I. Nikolaenko, *Deiatel'nost' KPSS po vnedreniiu v promyshlennost' dostizhenii nauki i peredovogo opyta* ('Vishcha shkola', Kiev, 1983) pp. 55–6; *Voprosy istorii KPSS*, 7/84, pp. 97–109.
117. *Pravda Ukrainy*, 24 November 1983, p. 1. Previously the council was headed by a deputy chairman of the Ukrainian Council of Ministers, but it was put under party control after evident dissatisfaction with its work. *Pravda Ukrainy*, 26 January 1977, p. 1.

CHAPTER 4

1. R. P. Suttmeier, 'Socialist science policy', *Problems of Communism*, vol. 26, no. 4 (July–August 1977) p. 75.
2. Medvedev (1969) op. cit., p. 56.
3. *Politichesky dnevnik 1964–70*, op. cit., pp. 303–6; Azbel, op. cit., p. 209.
4. *Partiinaia zhizn'*, 12/79, p. 59.
5. *Partiinaia zhizn*, 7/80, p. 37.
6. *Partiinaia zhizn'*, 12/79, p. 60.
7. Sharenkov, op. cit., p. 21; Z. A. Kozlova, *V tvorcheskom poiske. O rabote partiinykh organizatsii po povysheniiu trudovoi i obshchestvenno-politicheskoi aktivnosti nauchnykh kollektivakh v osushchestvlenii nauchno-tekhnicheskogo progressa* ('Moskovsky rabochy', Moscow, 1980) pp. 166–7.
8. *Kommunist Sovetskoi Latvii*, 1/61, pp. 52–3.
9. *VAN*, 1/64, pp. 45–6.
10. *VAN*, 4/65, p. 10.
11. *VAN*, 4/80, p. 43.
12. For further evidence, see *Voprosy filosofii*, 6/75, p. 160; *Sovetskaia Rossiia*, 28 August 1970, p. 2.
13. Turchin, op. cit., p. 12.
14. *Sovetskaia Rossiia*, 28 August 1970, p. 2.
15. *VAN*, 11/78, pp. 96–103.
16. *Marksistsko-leninskoe obrazovanie i nauchnoe tvorchestvo*, op. cit., pp. 139–44.
17. *Metodologiia nauki i nauchny progress* ('Nauka', Novosibirsk, 1981); Kugel', Lebin, Meleshchenko, op. cit., p. 146; *Ideino-politicheskoe vospitanie*, op. cit., p. 131; *Voprosy filosofii*, 6/75, p. 161.
18. *Metodologicheskie problemy kompleksnykh issledovanii*, op. cit.
19. *VAN*, 4/80, pp. 42, 49; *Voprosy filosofii*, 6/75, p. 158.
20. Ianovsky (1974) op. cit, p. 55; *VAN*, 4/80, pp. 44–5; Bakunin, op. cit., p. 132.
21. Ianovsky (1974) op. cit., p. 48.
22. *Kommunist*, 16/79, p. 42.
23. *VAN*, 4/80, p. 49.
24. *VAN*, 11/84, p. 53. One can even find a strange reference to the party buro of a methodological seminar in 1975. One can understand a semi-

nar perhaps having a party group, but a party buro sounds most unusual. *Voprosy filosofii*, 6/75, p. 160.

25. Ianovsky (1974) op. cit., p. 55. Regional party organs are still involved in the supervision of seminars. See *Ideino-politicheskoe vospitanie*, op. cit., p. 8; *Voprosy filosofii*, 6/75, p. 158.
26. *Ideino-politicheskoe vospitanie*, op. cit. p. 62.
27. See above, pp. 43–4.
28. *Partiinaia zhizn'*, 5/55, pp. 53–7.
29. for details, see J. W. Cleary, *Politics and Administration in Soviet Kazkahstan, 1955–1964*, Ph.D. dissertation, Australian National University, Canberra, 1967, p. 333; Gustafson (1981) op. cit., p. 30.
30. See above, pp. 26–31.
31. *Partiinaia zhizn'*, 19/69, p. 46.
32. S. Fortescue, 'Research institute party organizations and the right of control', *Soviet Studies*, vol. 35, no. 2 (April 1983) pp. 175–95. See also Kneen (1984) op. cit., ch. 7.
33. *Pravda Ukrainy*, 17 March 1977, p. 1; *Ideino-politicheskoe vospitanie*, op. cit., p. 155.
34. For another example, see Sharenkov, op. cit., p. 60.
35. See the party buro of a Ukrainian drafting bureau turning to the PPOs of enterprises to get essential supplies. *Pravda Ukrainy*, 26 January 1972, p. 2.
36. See Nekrich, op. cit.
37. *Partiinaia zhizn'*, 22/71, p. 30; 10/72, p. 42.
38. *Partiiny kontrol' deiatel'nosti administratsii* (Politizdat, Moscow, 1973) pp. 45–6.
39. See *Pravda Ukrainy*, 12 Febraury 1972, p. 2; *VAN*, 5/75, p. 38; *Partiinaia zhizn*, 24/80, p. 42; B. Harasymiw, *Political Elite Recruitment in the Soviet Union* (Macmillan, London and Basingstoke, 1984) pp. 170–1; *Partiinaia zhizn'*, 14/80, p. 19.
40. M. I. Khaldeev, *Pervichnaia partiinaia organizatsiia. Opyt, formy i metody raboty* (Politizdat, Moscow, 1974) p. 234.
41. For some examples of this phenomenon, see Nekrich, op. cit., pp. 77–81, 152, 164; Popovsky (1978) op. cit., pp. 64–5.
42. For some more details on these schemes, see Fortescue, *The Academy Reorganized*, op. cit., pp. 73–9.
43. See, for example, the purge in the Central Steam Turbine Institute in Leningrad, which was based on the new attestation procedures set out in the September 1968 Central Committee and Council of Ministers decree and led to the 'transfer to other work' of about 100 people. *Kommunist*, 18/68, p. 43. See also the cases of Burdzhalov and Mel'chuk. Nekrich, op. cit., p. 147; 'The case of Igor' Mel'chuk', *Survey*, vol. 23, no. 2 (103) (Spring 1977–78) pp. 126–40. Zhores Medvedev has described the effect of a negative *kharakteristika*. Medevedev (1972) op. cit., pp. 534–7.
44. *VAN*, 11/75, p. 16.
45. Expulsion, which requires a two-thirds majority of members attending the relevant PPO meeting, must be confirmed by the superior *raikom* or *gorkom* (as must reprimands recorded on the party card). *Spravochnik* (1980) op. cit., pp. 228–9.

46. Some rough-and-ready calculations and scattered Soviet data suggest that at least 20 per cent of staff of Soviet research institutes are party members, with a roughly similar percentage for research staff. One suspects that membership among institute directors is approaching universal, while rates among middle-level personnel would be around 70 per cent and rising. See Fortescue, 'Party membership', op. cit., pp. 144–51. See also Kneen (1984) op. cit., pp. 72–81.

47. *Pravda*, 21 March 1939, p. 6; 4 April 1939, p. 2; *Bol'shevik*, 4/39, pp. 46–7; *Partiinoe stroitel'stvo*, 4/39, p. 26.

48. For example, in 1973 the secretary of the party buro of the Kalinin *oblast'* clinical hospital said that his PPO had had no trouble handling the right of control since it had exercised control of the activity of the administration before 1971. *Kommunist*, 4/73, pp. 26–30.

49. See, for example, *Partiinaia zhizn'*, 22/71, pp. 28–30; *Partiiny kontrol'*, (1973) op. cit., p. 98.

50. The Decree itself appeared suddenly in the middle of a Central Committee plenum, although it was not mentioned on the agenda or in the discussions of the plenum. *KPSS v rezoliutsiiakh i resheniiakh s"ezdov, konferentsii i plenumov TsK*, pt 4, 1954–60, op. cit., pp. 527–8. It is presumably no coincidence that the plenum was concerned primarily with technological development of industry.

51. The only article of the time giving any details of their activities was written by the party secretary of a Rostov machine-building institute. *Partiinaia zhizn'*, 20/59, pp. 43–6. For opposition to the commissions, see *Partiinaia zhizn'*, 12/60, pp. 26–7. A recent publication claims that from 1959 commissions 'were set up not only in party organisations mentioned in the Instructions, but also in many others which received the right of control at the 24th Party Congress.' *Partiiny kontrol' deiatel'nosti administratsii* (Politizdat, Moscow, 1983) 2nd edition, p. 30.

52. *Party Rules*, 1961, para. 59.

53. *Partiinaia zhizn'*, 16/63, p. 33.

54. *Pravda Ukrainy*, 26 January 1972, p. 2; 6 February 1972, p. 2; Ianovsky (1979) op. cit., p. 208.

55. *Partiinaia zhizn'*, 16/63, p. 36.

56. R. G. Ianovsky, 'Rabota s nauchnoi intelligentsii' in *Partiia i intelligentsiia v usloviiakh razvitogo sotsializma* ('Mysl'', Moscow, 1977) p. 183.

57. *Partiinaia zhizn'*, 19/66, p. 21; 7/73, p 5; *Kommunist*, 11/72, p. 57.

58. *Partiinaia zhizn'*, 14/65, pp. 23–24; 10/68, pp. 41–2.

59. For example, *Kommunist Estonii*, 8/73, p. 2; For an exception, see Sliusarenko, op. cit., p. 87.

60. *Partiinaia zhizn'*, 2/77, p. 45; 14/77, p. 69.

61. *Moskovskaia pravda*, 17 September 1971, p. 2.

62. I have found three examples of reports being heard from individuals specifically recognised as not being party members. One refers to *soobshcheniia*, one to *doklady* and *soobshcheniia*, and one to *otchet*. *Partiinaia zhizn'*, 20/67, pp. 56–7; *Kommunist Estonii*, 11/71, p. 34.

63. *Spravochnik partiinogo rabotnika*, vyp. 23 (Politizdat, Moscow, 1983) pp. 487–92.

64. *Partiinaia zhizn'*, 9/77, p. 34; Moiseishin, op. cit., p. 101.

65. *Pravda Vostoka*, 1 April 1972, p. 2.
66. *Kommunist*, 13/64, p. 124.
67. *Kommunist Ukrainy*, 2/72, p. 67; *Partiinaia zhizn'*, 17/71, p. 18.
68. Moiseishin, op. cit., p. 101.
69. For an introduction to the long and complicated history of the people's control system, see G. Hodnett, 'Khrushchev and party-state control', in A. Dallin and A. F. Westin (eds), *Politics in the Soviet Union. Seven cases* (Harcourt, Brace and World, New York, 1966) ch. 4.
70. *Partiinaia zhizn'*, 2/69, p. 14; *Partiinaia zhizn' Kazakhstana*, 5/63, p. 26; 10/65, p. 72.
71. It would appear to be rare for a control group to examine the activity of the head of the institution, while the groups' directions (*ukazaniia*) appear to have only the force of recommendations.
72. *Partiinaia zhizn'*, 9/63, p. 19. See also the relevant chapters in Sisnev and Stepanov, op. cit.
73. *Partiinaia zhizn'*, 12/73 pp. 42–3; *Kommunist Ukrainy*, 6/71, p. 85; *Moskovskaia pravda*, 18 February 1971, p. 2.
74. *Kommunist Sovetskoi Latvii*, 7/70, p 79.
75. *Turkmenskaia iskra*, 12 September 1972, p. 2.
76. *Partiinaia zhizn'*, 19/66, p. 21; *Kommunist*, 11/72, p. 57. One authoritative statement, unusually, goes so far as to say that 'a manager cannot not consider the opinion of the party organization. Its decisions, suggestions and recommendations, if they are in complete agreement with party documents and Soviet law, are obligatory for him.' *Partiinaia zhizn'*, 17/77, p. 61.
77. *Partiinaia zhizn'*, 1/55, pp. 67–70.
78. *Partiinaia zhizn'*, 12/71, p. 36.
79. It is stated, of a hypothetical case: 'If communists in a *sovkhoz* see, for example, that decisions taken by the party and government on the development of agricultural production are not being adequately implemented, they have the right to get an explanation from the manager, work out their collective opinion on the measures needed to improve the situation, and in circumstances that require it to oblige (*obiazat'*) the director of the *sovkhoz* to put the measures into effect, naturally in strict accordance with party documents and Soviet laws.' *Partiinaia zhizn*, 14/65, p. 24.
80. *Partiinaia zhizn*, 17/78, p. 45.
81. *Pravda Ukrainy*, 25 February 1972, p. 2.
82. There are a number of words that are used by PPOs in their decisions requiring action from management. The strongest is 'to oblige' (*obiazat*), the weakest 'to ask' (*prosit'*). Somewhere in between are 'to suggest' (*predlozhit'*) and 'to demand' (*trebovat'* or *poruchit'*). Soviet commentators are often at pains to point out that the right of control should not be understood in terms of the different words PPOs can use in their decisions. *Uchitel'skaia gazeta* complained in 1973 that some people had the wrong idea of control. They believed that it gave the party organisation the right to 'suggest' measures to the director, whereas previously they could only 'ask'. Without actually saying that this view is incorrect, the paper stresses that the basic question is not the form of the decisions but the content, the essence. *Uchitel'skaia*

gazeta, 13 January 1973, p. 1. However some people who should know seem to think that the essence is precisely in those words. When I asked the former secretary of a ministry PPO what the right of control meant, he replied that it meant that a PPO could now 'suggest' as against 'ask' a manager to take action. When I asked what the difference is between the two words, he replied that a 'suggestion' meant that a certain period would be set aside for fulfilment of the PPO's recommendation. If the 'suggestion' was not implemented in this time the manager could have party disciplinary action taken against him.

83. *Partiinaia zhizn'*, 9/78, pp. 72–3.
84. For the exception which proves the rule, see the refusal of Khvostov, the director of the Institute of History, to take his place on that institute's 'liberal' party committee in the mid-1960s. Nekrich, op. cit., p. 250.
85. *Partiinaia zhizn'*, 5/55, p. 80.
86. *Partiinaia zhizn'*, 13/63, p. 52.
87. *Partiinaia zhizn'*, 22/74, pp. 45–7. In any case such as this one cannot be sure that the criticisms were not made at the instigation of the *raikom*.
88. *Partiinai zhizn'*, 10/68, pp. 41–2.
89. There is some confusion over who must appeal to higher authorities in the case of dispute. Some reports make PPO decisions obligatory for managers, it being up to the latter to appeal if there is disagreement. *Partiinaia zhizn'*, 23/65, p. 34; 17/77, p. 61. However most make it the responsibility of the PPO to appeal. *Partiinaia zhizn'*, 4/68, pp.40–1; 15/81, p. 67. This is what one would expect given the place of the PPO in the system.
90. *Partiinaia zhizn'*, 21/76, p. 72; *Izvestiia*, 28 July 1965, p. 3.
91. V. A. Rassudovsky, *Gosudarstvennaia organizatsiia nauki v SSSR* ('Iuridicheskaia literatura', Moscow, 1971) p. 133; *Literaturnaia gazeta*, 16 August 1972, p. 12; *Izvestiia*, 26 August 1969, p. 3. Since 1969 it has lost its power to *VAK* to determine the acceptance of dissertations.
92. Although see *Zaria Vostoka*, 8 April 1979, p. 2.
93. 'The case of Igor' Mel'chuk', op. cit., p. 127; Berg, op. cit., p. 309; *Kommunist (Vilnius)*, 2/70, p. 58; *Kommunist*, 6/72, p. 94.
94. *Pravda*, 15 May 1972, p. 2; *Kommunist (Vilnius)*, 2/70, p. 58; *Pravda*, 4 December 1972, p. 2; 'The case of Igor' Mel'chuk', op. cit., p. 126.
95. *Partiinaia zhizn'*, 16/63, pp. 38–9; 21/76, p. 72.
96. *Partiinaia zhizn' Kazakhstana*, 6/67, p. 44; *Pravda*, 24 Febraury 1972, p. 2.
97. *Moskovskaia pravda*, 10 September 1971, p. 2; *Kommunist Ukrainy*, 2/72, p. 66; *Pravda Vostoka*, 1 April 1972, p. 2.
98. In which cases party control is not always water-tight. See 'The case of Igor' Mel'chuk', op. cit. For other cases, see Nekrich, op. cit., p. 147; Berg, op. cit., pp. 309–23.
99. *Kommunist Moldavii*, 2/63, pp. 78–79.
100. *Partiiny kontrol'* (1983) op. cit., p. 48.
101. Medvedev (1969) op. cit., p. 53.
102. *Partiinaia zhizn*, 5/55, pp. 53–7.
103. *Kommunist*, 17/57, p. 41.
104. *Kommunist Sovetskoi Latvii*, 1/60, p. 13.
105. Solzhenitsyn, op. cit., p. 161.

106. Medvedev (1972) op. cit., p. 587.
107. See above, p. 28.
108. *Politichesky dnevnik 1964–70*, op. cit., pp. 303–6; B. Shragin, 'Oppozitsionnye nastroeniia v nauchnykh gorodakh', *SSSR. Vnutrennye protivorechiia*, no. 1 (1981) pp. 110–11.
109. See the Committee for Party Control reports in *Partiinaia zhizn'*, 13/70, p. 39; 17/74, p. 58; 11/77, p. 59; 8/78, p. 74; 17/78, p. 43.
110. See, for example, *Partiinaia zhizn'*, 19/69, p. 46; 18/84, pp. 31–2.
111. *Partiinaia zhizn'*, 19/59, p. 35.
112. *Kommunist*, 18/54, pp. 58–9; *Partiinaia zhizn' Kazakhstana*, 3/62, pp. 52–3; *Pravda*, 11 February 1974.
113. *Pravda Ukrainy*, 12 February 1972, p. 2.
114. *Partiinaia zhizn'*, 16/74, p. 53.
115. *Partiinaia zhizn'*, 12/76, p. 41.
116. For some examples, see *Partiinaia zhizn'*, 1/74, p. 47; *Kommunist Estonii*, 4/71, p. 30.
117. *Pravda*, 1 September 1983, p. 2.
118. *Pravda*, 10 January 1972, p. 2; 28 March 1972, p. 2.
119. See also the comments of a *raikom* secretary in *Pravda*, 14 June 1981, p. 2.
120. *Pravda*, December 1982, p. 2.
121. It should be repeated here that it is rare to see criticism of the chairman or director being expressed at such meetings. He, seemingly, can be dealt with only at higher levels.
122. *Partiinaia zhizn'*, 11/75, p. 71.
123. For an example of a research manager, a deputy director of the Kazakh Institute of Land Use, being protected by the *raikom* after the PPO had applied party disciplinary measures, see *Partiinaia zhizn'*, 8/55, p. 80.
124. *Ideino-politicheskoe vospitanie*, op. cit., p. 66.
125. Shcherbakov, op. cit., p. 98.
126. In this section I rely heavily on a previously published article, 'Party secretaries in Soviet research institutes: divided loyalties?', *Politics* (Australia), vol. 18, no. 1 (May 1983) pp. 73–83. Since the article is published in a journal that might not be readily available to some readers, I will reproduce much of it here, especially pp. 77–82.
127. *Ustav KPSS*, 1978, para. 57.
128. *Spravochnik sekretaria pervichnoi partiinoi organizatsii* (Politizdat, Moscow, 1980) 2nd edition, p. 126.
129. Ibid., pp. 129–32.
130. Ibid., p. 130.
131. *Kommunist Belorussii*, 10/70, p. 25.
132. *Partiinaia zhizn'*, 20/56, pp. 61–3.
133. *Partiinaia zhizn'*, 21/63, p. 49; 24/78, p. 64.
134. Nekrich, op. cit., pp. 249–52, 259–60.
135. The responses reported in the Miller survey sound reasonably close to what one would expect, although the relative weakness of the *raikom* in influencing the membership of the party buro or committee compared to the director is slightly surprising. Miller (1985) op. cit., p. 40.

136. I have used the name index of *Letopis' zhurnal'nykh statei* to trace articles and the biographical data hopefully accompanying them of people already identified as PPO secretaries. For details, see S. Fortescue, *The Primary Organisations of the Soviet Communist Party in Non-Production Institutions*, Ph.D. dissertation, Australian National University, Canberra, 1977, pp. 396–486.
137. In recent years the average size of research PPOs has been in the 90s, but one suspects that the median size is somewhat lower. Fortescue, 'Party membership', op. cit., p. 143.
138. Miller (1985) op. cit., p. 36.
139. Nekrich, op. cit., pp. 251–2.
140. The difference shows up in the Miller survey. While only 58 per cent of *vuz* PPO secretaries were engineers, 74 per cent and 86 per cent of Academy and branch secretaries were. Miller (1985) op. cit., p. 38.
141. Kozlova, op. cit., pp. 114, 169.
142. *Izvestiia*, 13 December 1984, p. 3.
143. Nekrich, op. cit., p. 249.
144. Ibid., p. 307.
145. *Spravochnik* (1980) op. cit., p. 130.
146. Ibid., pp. 130–1; *Partiinaia zhizn'*, 20/81, p. 74.
147. *Partiinaia zhizn'*, 12/64, pp. 44–5.
148. The Miller survey finds that on the whole research PPO secretaries are seen by their colleagues as at best average scholars, although Academy researchers tend to give them higher ratings. Miller (1985) op. cit., p. 47.
149. M. Agursky, *The Research Institute of Machine-Building Technology. A part of the Soviet military-industrial complex*, Soviet Institutions Series, Paper No. 8, The Hebrew University of Jerusalem, Jerusalem, 1976, p. 51.
150. Popovsky (1978) op. cit., pp. 55–8.
151. Fortescue (1977) op. cit., pp. 470–6.
152. Ibid., pp. 438–41, 470–6.

CONCLUSION

1. Miller, (1985) op. cit., p. 57.
2. Gustafson (1981) op. cit., p. 143.
3. Hough in Solomon, op. cit., pp. 43, 51–52.
4. Gustafson (1981) op. cit., p. 158; E. P. Hoffmann, 'Technology, values and political power in the Soviet Union: do computers matter?', in Fleron, *Technology and Communist Culture*, op. cit., p. 409; Hodnett in Bialer, op. cit., p. 108.
5. Solomon, P. H. (1978) op. cit.
6. Gustafson (1981) op. cit., p. 157; Hoffmann in Fleron, op. cit., p. 413.
7. For example, Gustafson (1981) op. cit; Schwartz and Keech, op. cit.
8. Fleron in Farrell, op. cit., pp. 118–20.
9. Wood in Gilpin and Wright, op. cit.
10. Hodnett in Bialer, op. cit., p. 91.

Bibliography

ADAMS, M. B. 'Science, ideology, and structure: the Kol'tsov institute, 1900–1970', in L. L. Lubrano and S. G. Solomon (eds), *The Social Context of Soviet Science*, Westview, Boulder and Dawson, Folkestone, 1980.

AGURSKY, M. *The Research Institute of Machine-Building Technology. A part of the Soviet military–industrial complex*. Soviet Institutions Series, Paper No. 8, The Hebrew University of Jerusalem, Jerusalem, 1976.

AMAL'RIK, A. *Zapiski dissidenta*. Ardis, Ann Arbor, 1982.

AMANN, R. 'The Soviet research and development system. The pressure of academic tradition and rapid industrialization', *Minerva*, vol. 8, no. 2, April 1970, pp. 217–41.

AMANN, R. and COOPER, J. (eds), *Industrial Innovation in the Soviet Union*. Yale UP, New Haven and London, 1982.

ANASTASYN, D. and VOZNESENSKY, I., 'Nachalo trekh natsional'nykh akademii', *Pamiat'. Istorichesky sbornik*, vyp. 5, 1982, pp. 165–225.

ANDRONIKASHVILI, E. *Nachinaiu s El'brusa (Tvorcheskie portrety uchenykh)*. 'Metsnieraba', Tbilisi, 1982.

ASHBY, E. *Scientist in Russia*. Penguin, Harmondsworth, 1947.

AZBEL, M. Ya. *Refusenik. Trapped in the Soviet Union*. Hamish Hamilton, London, 1981.

BAILES, K. E. 'The politics of technology. Stalin and technocratic thinking among Soviet engineers', *American Historical Review*, vol. 79, no. 2, April 1974, pp. 445–69.

BAKUNIN, A. V. (ed.). *Deiatel'nost' KPSS po uskoreniiu nauchno-tekhnicheskogo progressa*. 'Vysshaia shkola', Moscow, 1980.

BALZER, H. D. 'The Soviet scientific and technical intelligentsia', *Problems of Communism*, vol. 31, no. 3, May–June 1982, pp. 66–72.

BARMIN, I. P. *Iz opyta raboty KPSS i sovetskogo gosudarstva po sozdaniiu kadrov sovetskoi intelligentsii (1928–1933)*. Izd-vo Moskovskogo un-ta, Moscow, 1965.

BARRINGTON MOORE. *Terror and Progress USSR. Some sources of change and stability in the Soviet dictatorship*. Harvard UP, Cambridge Mass, 1954.

BELIAEV, E. A. *KPSS i organizatsiia nauki v SSSR*. Politizdat, Moscow, 1982.

BELIAKOV, V. K. and ZOLOTAREV, N. A. *Organizatsiia udesiateriaet silu. Razvitie organizatsionnoi struktury KPSS. 1917–1974gg*. Politizdat, Moscow, 1975.

BELITSKY, E. *The Central Institute of Mathematical Economics*. Soviet

Institutions Series, Paper No. 1, The Hebrew University of Jerusalem, Jerusalem, 1974.

BELOUSOV, R. A. and KHROMUSHIN, G. B. (eds). *Partiinoe rukovodstvo ekonomikoi.* 'Ekonomika', Moscow, 1981.

BERG, R. *Sukhovei. Vospominaniia genetika.* Chalidze, New York, 1983.

BIRMAN, I. 'Protivorechivye protivorechiia (zametki o sovetskoi ekonomicheskoi teorii i praktiki', *SSSR. Vnutrennye protivorechiia,* no. 1, 1981, pp. 9–63.

BESANCON, A. 'Solzhenitsyn at Harvard', *Survey,* vol. 24, no. 1 (106), Winter 1979, pp. 133–44.

BLIAKHMAN, L. S. and SHKARATAN, O. I. *NTR, rabochy klass, intelligentsiia.* Politizdat, Moscow, 1973.

BORCKE, A. von. 'Dilemmas of single-party rule in the Soviet Union', *Problems of Communism,* vol. 31, no. 3, May–June 1982, pp. 60–5.

BOTTOMORE, T. and RUBEL, M. (eds). *Karl Marx. Selected writings in sociology and social philosophy.* Penguin, Harmondsworth, 1967.

BRESLAUER, G. W. *Khrushchev and Brezhnev as Leaders. Building authority in Soviet politics.* Allen and Unwin, London, 1982.

BREZHNEV, L. I. *Leninskim kursom: rechi i stat'i.* Politizdat, Moscow, 1970–82, 9 volumes.

BREZHNEV, L. I. *Tselina.* Politizdat, Moscow, 1978.

BROWN, A. 'Andropov: discipline *and* reform?', *Problems of Communism,* vol. 32, no. 1, January–February 1983, pp. 18–31.

BROWN, A. 'Pluralism, power and the Soviet political system: a comparative perspective', in S. G. Solomon (ed.), *Pluralism in the Soviet Union. Essays in honour of H. Gordon Skilling.* Macmillan, London and Basingstoke, 1983, ch. 4.

BROWN, A. 'Political science in the Soviet Union: a new stage of development?', *Soviet Studies,* vol. 36, no. 3, July 1984, pp. 317–44.

BRUMAS, M. 'USSR discriminates against Jews in higher education', *Radio Liberty,* RL 301/84, 6 August 1984.

BULGANIN, N. A. *O zadachakh po dal'neishemu pod''emu promyshlennosti, tekhnicheskomu progressu i uluchsheniiu organizatsii proizvodstva.* Gospolitizdat, Moscow, 1955.

BUNCE, V. 'The political economy of the Brezhnev era. The rise and fall of corporatism', *British Journal of Political Science,* vol. 13, pt 2, April 1983, pp. 129–58.

BUNCE, V. and ECHOLS, J. M., III. 'Soviet politics in the Brezhnev era: "pluralism" or "corporatism" ', in D. R. Kelly (ed.), *Soviet Politics in the Brezhnev Era.* Praeger, New York, 1980, pp. 1–26.

'The case of Igor' Mel'chuk', *Survey,* vol. 23, no. 2 (103), Spring 1977–8, pp. 126–40.

CHEKHARIN, E. *Rol' nauki v sozdanii nauchno-tekhnicheskoi bazy kommunizma.* 'Znanie', Moscow, 1965.

CHEKHARIN, E. *Sovetskaia politicheskaia sistema v usloviiakh razvitogo sotsializma.* 'Mysl'', Moscow, 1975.

CHEMODANOV, M. *Prevrashchenie nauki v neposredstvennuiu proizvodstvennuiu silu v protsesse stroitel'stva kommunizma,* 'Mysl'', Moscow, 1966.

CHOTINER, B. A. *Khrushchev's Party Reform. Coalition building and institutional innovation.* Greenwood, Westport, Conn., 1984.

CLEARY, J. W. *Politics and Administration in Soviet Kazakhstan, 1955–1964,* Ph.D. dissertation, Australian National University, Canberra, 1967.

CLEMENS, W. C., Jr. 'Alexander Yanov – writing between the lines', *Problems of Communism*, vol. 27, no. 3, May–June 1978, pp. 65–8.

COCKS, P., DANIELS, R. V. and HEER, N. W. (eds). *The Dynamics of Soviet Politics.* Harvard UP, Cambridge, Mass., and London, 1976.

CONNOR, W. D. 'Differentiation, integration, and political dissent in the USSR', in R. L. TOKES (ed.), *Dissent in the USSR. Politics, ideology and people.* Johns Hopkins UP, Baltimore and London, 1975, ch. 4.

CONQUEST, R. *The Politics of Ideas in the USSR.* Bodley Head, London, 1967.

COOPER, J. M. 'The scientific and technical revolution in Soviet theory', in F. J. Fleron (ed.), *Technology and Communist Culture. The socio-cultural impact of technology under socialism.* Praeger, New York, 1977, ch. 3.

COOPER, J. M. 'The scientific and technical revolution in the USSR', paper presented at the NASEES Annual Conference, Cambridge, 1981.

DAWISHA, K. 'The limits of the bureaucratic politics model: observations on the Soviet case', *Studies in Comparative Communism*, vol. 13, no. 4, Winter 1980, pp. 300–46.

Deputaty Verkhovnogo Soveta SSSR. Deviaty sozyv. Moscow, 1979.

Directory of Soviet Officials. National organizations. A reference aid. Directorate of Intelligence, Washington DC, 1979.

DOBROV, G. M. *Nauka o nauke.* 'Naukova dumka', Kiev, 1970.

DOROFEEV, V. and V. *Vremia, uchenye, sversheniia.* Politizdat, Moscow, 1975.

DUMACHEV, A. P. *Khozraschetnye ob"edineniia v promyshlennosti.* Lenizdat, Leningrad, 1972.

DUMACHEV, A. P. *Partiinye organizatsii i proizvodstvennye ob"edineniia.* Politizdat, Moscow, 1976.

DUMACHEV, A. P. and SOLOV'EV, A. P. *Za effektivnost' nauki.* Lenizdat, Leningrad, 1968.

DUNMORE, T. *The Stalinist Command Economy. The Soviet state apparatus and economic policy, 1943–53.* Macmillan, London and Basingstoke, 1980.

DUZHENKOV, V. I. *Problemy organizatsii nauki (Regional'nye aspekty).* 'Nauka', Moscow, 1978.

XXIII s"ezd Kommunisticheskoi Partii Sovetskogo Soiuza. Stenografichesky otchet. Politizdat, Moscow, 1966.

ELLMAN, M. *Soviet Planning Today. Proposals for an optimally functioning economic system.* Cambridge UP, 1971.

ELLUL, J. *The Technological Society.* Cape, London, 1965.

EVANS, A. 'Developed socialism in Soviet ideology', *Soviet Studies*, vol. 29, no. 3, July 1977, pp. 409–28.

FAINSOD, M. *How Russia is Ruled.* Harvard UP, Cambridge, Mass., 1967, revised edition.

FEDOSEEV, A. *Zapadnia. Chelovek i sotsializm.* Posev, Frankfurt/Main, 1976.

FITZPATRICK, S. *Education and Social Mobility in the Soviet Union,* Cambridge UP, 1979.

FLERON, F. J. 'Co-optation as a mechanism of adaption to change: the Soviet political leadership system', in R. E. Kanet (ed.), *The Behavioral Revolution and Communist Studies.* Free Press, New York, 1971, pp. 125–49.

FLERON, F. J. 'Representation of career types in the Soviet political leadership', in R. B. Farrell (ed.), *Political Leadership in Eastern Europe and the Soviet Union.* Aldine, Chicago, 1970, ch. 6.

FLERON, F. J. (ed.) *Technology and Communist Culture. The socio-cultural impact of technology under socialism.* Praeger, New York, 1977.

FORTESCUE, S. *The Academy Reorganized. The R&D role of the Soviet Academy of Sciences since 1961,* Occasional paper No. 17, Department of Political Science, RSSS, Australian National University, Canberra, 1983.

FORTESCUE, S. 'Akademiia nauk –gotovnost' k kompromissu', *Obozrenie,* no. 10, July 1984, pp. 7–10.

FORTESCUE, S. 'Party membership in Soviet research institutes', *Soviet Union/Union Sovietique,* vol. 11, pt 2, 1984, pp. 129–56.

FORTESCUE, S. 'Party secretaries in Soviet research institutes: divided loyalties?' *Politics* (Australia), vol. 18, no. 1, May 1983, pp. 73–83.

FORTESCUE, S. *The Primary Organisations of the Soviet Communist Party in Non-Production Institutions,* Ph.D. dissertation, Australian National University, Canberra, 1977.

FORTESCUE, S. 'Project planning in Soviet R&D', *Research Policy,* vol. 14 (1985) pp. 267–82.

FORTESCUE, S. 'Research institute party organizations and the right of control', *Soviet Studies,* vol. 35, no. 2, April 1983, pp. 175–95.

FORTESCUE, S. 'A Soviet interest system', Department of Political Science, RSSS, Australian National University, Canberra, April 1974 (unpublished).

FRANK, A. G. 'Goal ambiguity and conficting standards: an approach to the study of organization', *Human Organization,* vol. 17, no. 4, Winter 1958–59, pp. 8–13.

FRANKS, F. *Polywater,* MIT Press, Cambridge, Mass., and London, 1981.

FRIEDRICH, C. J. and BRZEZINSKI, Z. K. *Totalitarian Dictatorship and Autocracy.* Praeger, New York, 1966, 2nd edition.

FEUER, L. 'The intelligentsia in opposition', *Problems of Communism,* vol. 19, no. 6, November–December 1970, pp. 1–16.

GALLAGHER, M. P. and SPIELMANN, K. F. *Soviet Decision-Making for Defense. A critique of US perspectives on the arms race.* Praeger, New York, 1972.

GARDER, M. *Die Agonie des Sowjetsystems.* Ullstein, 1965.

GEHLEN, M. P. 'Group theory and the study of Soviet politics', in S. I. Ploss (ed.), *The Soviet Political Process. Aims, techniques, and examples of analysis.* Ginn, Waltham, Mass, 1971.

GLAZOV, Iu. *Tesnye vrata. Vozrozhdenie russkoi intelligentsii.* Overseas Publications Interchange, London, 1973.

GLEBOVA, I. A. and SIGOV, I. I. *Sovershenstvovanie upravleniia fundamental'nymi issledovaniiami v krupnom gorode.* 'Nauka', Leningrad, 1983.

GOLOVIN, I. *Academician Igor Kurchatov.* 'Mir', Moscow, 1969.

GRAHAM, L. R. 'Biomedicine and the politics of science in the USSR', *Soviet Union/Union Soviétique,* vol. 8, pt 2, 1981, pp. 147–58.

GRAHAM, L. R. 'The development of science policy in the Soviet Union', in T. Dixon Long and C. Wright (eds). *Science Policies of Industrial Nations*. Praeger, New York, 1975, ch. 2.

GRAHAM, L. R. 'Quantum mechanics and dialectical materialism', *Slavic Review*, vol. 25, no. 3, Spring 1966, pp. 381–410.

GRAHAM, L. R. 'The role of the Academy of Sciences', *Survey*, vol. 23, no. 1 (102), Winter 1977–78, pp. 117–33.

GRAHAM, L. R. *Science and Philosophy in the Soviet Union*. Allen Lane, London, 1973.

GRAHAM, L. R. *The Soviet Academy of Sciences and the Communist Party, 1917 – 1932*. Princeton UP, New Jersey, 1967.

GREENBERG, L. L. 'Soviet science policy and the scientific establishment', *Survey*, vol. 17, no. 4 (81), Autumn 1971, pp. 51–63.

GUBAREV, V. *Konstruktor. Neskol'ko stranits iz zhizni Mikhaila Kuz'micha Iangelia*. Politizdat, Moscow, 1977.

GUROFF, G. 'The Red-expert debate. Continuities in the state-entrepreneur tension', in G. Guroff and F. V. Carstensen (eds), *Entrepreneurship in Imperial Russia and the Soviet Union*. Princeton UP, New Jersey, 1983.

GUSTAFSON, T. 'American science and Soviet concerns', *Survey*, vol. 23, no. 1 (102), Winter 1977–78, pp. 146–60.

GUSTAFSON, T. 'Environmental conflict in the USSR', in D. Nelkin (ed.), *Controversy. Politics of technical decisions*. Sage, London and Beverly Hills, 1979, ch. 3.

GUSTAFSON, T. *Reform in Soviet Politics. Lessons of recent policies on land and water*, Cambridge UP, New York, 1981.

GVAT', I· 'The creativity of F. M. Dostoevsky as a permanent provocation', *Radio Svoboda*, RS 138/83, 13 July 1983.

GVISHIANI, D. M. (ed.). *Voprosy teorii i praktiki upravleniia i organizatsii nauki*. 'Nauka', Moscow, 1975.

HAHN, J. W. 'The role of Soviet sociologists in the making of social policy', in R. B. Remnek (ed.), *Social Scientists and Policy Making in the USSR*. Praeger, New York, 1977, ch. 2.

HAHN, W. G. *Postwar Soviet Politics. The fall of Zhdanov and the defeat of moderation*. Cornell UP, Ithaca and London, 1982.

HAMMER, D. P. 'Brezhnev and the Communist Party', *Soviet Union/Union Soviétique*, vol. 2, pt 1, 1975, pp. 1–21.

HARASYMIW, B. *Political Elite Recruitment in the Soviet Union*. Macmillan, London and Basingstoke, 1984.

HELGESON, A. 'Recent developments on the Soviet sociological scene', *Radio Liberty*, RL 461/83, 8 December 1983.

HILL, R. J. *Soviet Politics, Political Science and Reform*. Martin Robertson, Oxford, 1980.

HILL, R. J., DUNMORE, T. and DAWISHA, K. 'The USSR. The revolution reversed?' in L. Holmes (ed.), *The Withering Away of the State. Party and state under communism*. Sage, London and Beverly Hills, 1981, ch. 9.

HILL, R. J. and FRANK, P. *The Soviet Communist Party*. Allen and Unwin, London, 1981.

HODNETT, G. 'Khrushchev and party-state control', in A. Dallin and A. F. Westin (eds), *Politics in the Soviet Union. Seven cases*. Harcourt, Brace and World, New York, 1966, ch. 4.

HODNETT, G. 'The pattern of leadership politics', in S. Bialer (ed.), *The Domestic Context of Soviet Foreign Policy*, Westview, Boulder, 1981, ch. 4.

HODNETT, G. and OGAREFF, V. *Leaders of the Soviet Republics, 1955–1972*, Department of Political Science, RSSS, Australian National University, Canberra, 1973.

HOFFMANN, E. P. 'Soviet views of "the STR" ', *World Politics*, vol. 30, no. 4, July 1978, pp. 615–44.

HOFFMANN, E. P. 'Technology, values and political power in the Soviet Union: do computers matter?' in F. J. Fleron (ed.), *Technology and Communist Culture. The socio-cultural impact of technology under socialism*. Praeger, New York, 1977, ch. 9.

HOFFMANN, E. P. and LAIRD, R. F. *The Scientific-Technological Revolution and Soviet Foreign Policy*. Pergamon, New York, 1982.

HOLLOWAY, D. *Entering the Nuclear Arms Race. The Soviet decision to build the atomic bomb, 1939–45*, International Security Studies Program Working Paper No. 9, Wilson Center, Washington DC, 1979.

HOLLOWAY, D. 'Innovation in science – the case of cybernetics in the Soviet Union', *Science Studies*, vol. 4, no. 4, October 1974, pp. 299–337.

HOSKIN, G. *Beyond Socialist Realism. Soviet fiction since Ivan Denisovich*. Granada, London, 1980.

HOUGH, J. F. 'The bureaucratic model and the nature of the Soviet system', *Journal of Comparative Administration*, vol. 5, no. 2, August 1973, pp. 134–67.

HOUGH, J. F. 'Pluralism, corporatism and the Soviet Union', in S. G. Solomon (ed.), *Pluralism in the Soviet Union. Essays in honour of H. Gordon Skilling*. Macmillan, London and Basingstoke, 1983, ch. 3.

HOUGH, J. F. 'The party apparatchiki', in H. G. Skilling and F. Griffiths (eds), *Interest Groups in Soviet Politics*, Princeton UP, New Jersey, 1971, ch. 3.

HOUGH, J. F. *The Soviet Prefects. The local party organs in industrial decision-making*, Harvard UP, Cambridge, Mass., 1969.

HOUGH, J. F. 'Soviet policy-making toward foreign communists', *Studies in Comparative Communism*, vol. 15, no. 3, Autumn 1982, pp. 167–83.

HOUGH, J. F. 'Soviet succession: issues and personalities', *Problems of Communism*, vol. 31. no. 5, September–October 1982, pp. 20–40.

HOUGH, J. F. 'The Soviet system. Petrification or pluralism?' *Problems of Communism*, no. 2, March–April 1972, pp. 25–45.

HOUGH, J. F. and FAINSOD, M. *How the Soviet Union is Governed*. Harvard UP, Cambridge, Mass., 1979.

IAKHOT, I. 'Sud'ba statistiki i materialisticheskikh metodov v sotsiologii (30-e gody)', *SSSR. Vnutrennye protivorechiia*, no. 1, 1981, pp. 158–200.

IANEVICH, N. 'Institut mirovoi literatury v 1930-e – 1970-e gody', *Pamiat'. Istorichesky sbornik*, no. 5, 1982, pp. 83–162.

IANOVSKY, R. G. *Formirovanie lichnosti uchenogo v usloviiakh razvitogo sotsializma*. 'Nauka', Novosibirsk, 1979.

IANOVSKY, R. G. *Politicheskaia ucheba v nauchnom kollektive*. Politizdat, Moscow, 1974.

IANOVSKY, R. G. 'Rabota s nauchnoi intelligentsii', in *Partiia i intelligentsiia v usloviiakh razvitogo sotsializma*. 'Mysl'', Moscow, 1977, ch. 9.

224   *Bibliography*

Ideino-politicheskoe vospitanie nauchno-tekhnicheskoi intelligentsii. 'Nauka', Moscow, 1982.

Istoriia Kommunisticheskoi Partii Sovetskogo Soiuza, vol. 5, bks 1 and 2. Politizdat, Moscow, 1970 and 1980.

IVANOVA, L. V. *Formirovanie sovetskoi nauchnoi intelligentsii (1917–27gg.).* 'Nauka', Moscow, 1980.

JONES, E. 'Committee decision making in the Soviet Union', *World Politics,* vol. 36, no. 2, January 1984, pp. 165–88.

JORAVSKY, D. 'Bosses and scientists', *Problems of Communism,* vol. 16, no. 1, January–Febraury 1967, pp. 72–5.

JORAVSKY, D. *The Lysenko Affair.* Harvard UP, Cambridge, Mass., 1970.

JORAVSKY, D. *Soviet Marxism and Natural Science, 1917–1932.* Columbia UP, New York, 1961.

JOSEPHSON, P. R. 'Science and ideology in the Soviet Union: the transformation of science into a direct productive force', *Soviet Union/Union Soviétique,* vol. 8, pt. 2, 1981, pp. 159–85.

JOZSA, G. 'Political *Seilschaften* in the USSR', in T. H. Rigby and B. Harasymiw (eds), *Leadership Selection and Patron–Client Relations in the USSR and Yugoslavia.* Allen and Unwin, London, 1983, ch. 4.

KPSS v rezoliutsiiakh i resheniiakh s"ezdov, konferentsii i plenumov TsK, pt 4, 1954–60, Politizdat, Moscow, 1960, 7th edition.

KPSS v rezoliutsiiakh i resheniiakh s"ezdov, konferentsii i plenumov TsK, vol. 11, 1972–75. Politizdat, Moscow, 1976.

KANYGIN, Iu. M. and BOTVIN, V. A. *Problemy razvitiia i ispol'zovaniia nauchnogo potentsiala krupnykh gorodov.* 'Naukova dumka', Kiev, 1980.

KATSENELINBOIGEN, A. 'Nuzhny li v SSSR Don Kikhoty?' *SSSR. Vnutrennye protivorechiia,* no. 2, 1981, pp. 248–52.

KATSENELINBOIGEN, A. 'Planning and Soviet science', *Survey,* vol. 23, no. 1 (102), Winter 1977–78, pp. 42–52.

KATSENELINBOIGEN, A. *Soviet Economic Thought and Political Power in the USSR.* Pergamon, New York, 1980.

KHALDEEV, M. I. *Pervichnaia partiinaia organizatsiia. Opyt, formy i metody raboty.* Politizdat, Moscow, 1974.

Khrushchev Remembers. The last testament. Deutsch, London, 1974.

KNEEN, P. 'Why natural scientists are a problem for the CPSU', *British Journal of Political Science,* vol. 8, pt 2, April 1978, pp. 177–98.

KNEEN, P. *Soviet Scientists and the State.* Macmillan, London and Basingstoke, 1984.

KOL'MAN, A. *My ne dolzhny byli tak zhit'.* Chalidze, New York, 1982.

KOL'TSOV, A. V. *Razvitie Akademii nauk kak vysshego nauchnogo uchrezhdeniia SSSR. 1926–1932.* 'Nauka', Leningrad, 1982.

KOMKOV, G. D., LEVSHIN, B. V. and SEMENOV, L. K. *Akademiia nauk SSSR. Kratky istorichesky ocherk.* 'Nauka', Moscow, 1977, 2nd edition, 2 vols.

KONING, J. W. *The Scientists Look at Research Management.* AMACOM, New York, 1976.

KOSOLAPOV, V. V. AND SHCHERBAN', A. N. *Optimizatsiia nauchno-issledovatel'skoi deiatel'nosti.* 'Naukova dumka', Kiev, 1971.

KOZLOV, B. I. (ed.). *Organizatsiia i razvitie otraslevykh nauchno-issledovatel'skikh institutov Leningrada. 1917–1977.* 'Nauka', Leningrad, 1979.

KOZLOVA, Z. A. *V tvorcheskom poiske. O rabote partiinykh organizatsii po povysheniiu trudovoi i obshchestvenno-politicheskoi aktivnosti nauchnykh kollektivakh v osushchestvlenii nauchno-tekhnicheskogo progressa.* 'Moskovsky rabochy', Moscow, 1980.

KRAUS, H. *The Composition of Leading Organs of the CPSU (1952–1976),* supplement to the Radio Liberty Research Bulletin, 1976.

KUGEL', S. A., LEBIN, B. D. and MELESHCHENKO, Iu. S. (eds). *Nauchnye kadry Leningrada. Struktura kadrov i sotsial'nye problemy organizatsii truda.* 'Nauka', Leningrad, 1973.

LAKHTIN, G. A. (ed.). *Osnovy upravleniia naukoi. Ekonomicheskie problemy,* 'Nauka', Moscow, 1983.

LANE, D. and O'DELL, F. *The Soviet Industrial Worker. Social class, education and control.* Martin Robertson, Oxford, 1978.

LANGSAM, D. E. and PAUL, D. W. 'Soviet politics and the group approach: a conceptual note', *Slavic Review,* vol. 31, no. 1, March 1972, pp. 136–41.

LAVIGNE, M. 'Advanced socialist society', *Economy and Society,* vol. 7, no. 4, November 1978, pp. 367–94.

LEMBRUCH, G. 'Interest intermediation in capitalist and socialist systems. Some structural and functional perspectives in comparative research', *International Political Science Review,* vol. 4, no. 2, 1983, pp. 153–72.

LEWIS, R. *Science and Industrialization in the USSR. Industrial research and development, 1917–1940.* Macmillan, London and Basingstoke, 1979.

LOWENHARDT, J. *Decision Making in Soviet Politics.* Macmillan, London and Basingstoke, 1981.

LOWENHARDT, J. *The Soviet Politburo.* Canongate, Edinburgh, 1982.

LOWENHARDT, J. ' "The tale of the torch". Scientist entrepreneurs in the Soviet Union', *Survey,* vol. 20, no. 4 (93), Autumn 1974, pp. 113–21.

LUBRANO, L. L 'National and international politics in US-USSR cooperation', *Social Studies of Science,* vol. 11, no. 4, November 1981, pp. 451–80.

LUBRANO, L. L. 'Soviet science specialists: professional roles and policy involvement', in R. B. Remnek (ed.), *Social Scientists and Policy Making in the USSR,* Praeger, New York, 1977, ch. 3.

LUBRANO, L. L. *Soviet Sociology of Science.* American Association for the Advancement of Slavic Studies, Columbus, Ohio, 1976.

LUBRANO, L. L. 'Survey research as a source of information for Soviet science policy', *Soviet Union/Union Soviétique,* vol. 9, pt 1, 1982, pp. 55–81.

MCCAGG, W. O. *Stalin Embattled, 1943–1948.* Wayne State UP, Detroit, 1978.

MCHALE, V. E. and MASTRO, J. P. 'The Central Committee of the CPSU', paper presented to the 66th Annual Meeting of the American Political Science Association, Los Angeles, September 1970.

Marksistsko-leninskoe obrazovanie i nauchnoe tvorchestvo. 'Nauka i tekhnika', Minsk, 1982.

MATTHEWS, M. *Class and Society in Soviet Russia.* Allen Lane, London, 1972.

MATTHEWS, M. *Education in the Soviet Union. Policies and institutions since Stalin.* Allen and Unwin, London, 1982.

MEDVEDEV, R. *Kniga o sotsialisticheskoi demokratii.* Alexander Herzen Foundation and Editions Grasset & Fasquelle, Amsterdam and Paris, 1972.

MEDVEDEV, R. *Let History Judge. The origins and consequences of Stalinism.* Macmillan, London, 1972.

MEDVEDEV, Z. A. *Mezhdunarodnoe sotrudnichestvo uchenykh i natsional'nye granitsy. Taina perepiski okhraniaetsia zakonom.* Macmillan, London and Basingstoke, 1972.

MEDVEDEV, Z. A. *Nuclear Disaster in the Urals.* Vintage, New York, 1980.

MEDVEDEV, Z. A. *The Rise and Fall of T. D. Lysenko.* Columbia UP, New York and London, 1969.

MEDVEDEV, Z. A. *Soviet Science*, Oxford UP, 1979.

MEISSNER, B. 'The Soviet Union and its area of hegemony and influence', in H. J. Veen et al., *Is Communism Changing?* Hase & Koehler, Mainz, 1981.

Metodologicheskie problemy kompleksnykh issledovanii. 'Nauka', Novosibirsk, 1983.

Metodologiia nauki i nauchny progress. 'Nauka', Novosibirsk, 1981.

MEYER, A. C. *The Soviet Political System: an interpretation.* Random House, New, York, 1965.

MEYER, A. C. 'USSR Incorporated', in D. W. Treadgold (ed.), *The Development of the USSR. An exchange of views.* University of Washington Press, Seattle, 1964, pp. 21–28.

MEYNAUD, J. *Technocracy.* Faber, London, 1968.

MIKULINSKY, S. R. and RIKHTA, R. (eds). *Sotsializm i nauka* 'Nauka', Moscow, 1981.

MILLER, R. F. 'The role of the Communist Party in Soviet research and development', *Soviet Studies*, vol. 37, no. 1, January 1985, pp. 31–59.

MILLER, R. F. 'The scientific–technical revolution and the Soviet administrative debate', in P. Cocks, R. V. Daniels and N. W. Heer (eds), *The Dynamics of Soviet Politics*, Harvard UP, Cambridge, Mass., and London, 1976, ch. 8.

MILLER, R. F. 'Soviet policy on science and technology and East–West relations', *World Review*, vol. 19, no. 2, June 1980, pp. 61–76.

MILLS, R. M. 'Lysenkoism: the un-science', *Problems of Communism*, vol. 21, no. 5, September–October, 1972, pp. 84–6.

MLYNAR, Z. *Night Frost in Prague.* Karz, New York, 1980.

MOISEISHIN, O. Kh. *Deistvennost' partiinogo kontrolia. Iz opyta Kompartii Belorussii.* 'Belarus', Minsk, 1982.

MOULY, R. W. 'Values and aspirations of Soviet youth', in P. Cocks, R. V. Daniels and N. W. Heer (eds). *The Dynamics of Soviet Politics.* Harvard UP, Cambridge, Mass., and London, 1976, ch. 12.

NARIN, F., DAVIDSON FRAME, J. and CARPENTER, M. P. 'Highly cited Soviet papers: an exploratory investigation', *Social Studies of Science*, vol. 13, 1983, pp. 307–19.

Nauchno-organizatsionnaia deiatel'nost' akademika A. F. Ioffe. Sbornik dokumentov. 'Nauka', Leningrad, 1980.

NEKRICH, A. *Otreshis' ot strakha. Vospominaniia istorika.* Overseas Publications Interchange, London, 1979.

NELIDOV, D. 'Ideokraticheskoe soznanie i lichnost", in *Samosoznanie. Sbornik statei.* 'Khronika', New York, 1976.

NIKOLAENKO, I. I. *Deiatel'nost' KPSS po vnedreniiu v promyshlennost' dostizhenii nauki i peredovogo opyta.* 'Vishcha shkola', Kiev, 1983.

OBERG, J. E. *Red Star in Orbit.* Harrap, London, 1981.

ODOM, W. E. 'A dissenting view on the group approach to Soviet politics', *World Politics*, vol. 28, no. 4, July 1976, pp. 542–67.

OGAREFF, V. *Leaders of the Soviet Republics. 1971–1980*, Department of Political Science, RSSS, Australian National University, Canberra, 1980.

ORLOVA, R. *Vospominaniia o neproshedshem vremeni.* Ardis, Ann Arbor, 1983.

PARROTT, B. B. *Politics and Technology in the Soviet Union.* MIT Press, Cambridge, Mass., and London, 1983.

PARROTT, B. B. *Technology and the Soviet Polity. The problem of industrial innovation, 1928–1973*, Ph.D. dissertation, Columbia University, New York, 1976.

PARRY, A. *The New Class Divided. Science and technology versus communism.* Macmillan, New York, 1966.

Partiiny kontrol' deiatel'nosti administratsii. Politizdat, Moscow, 1973.

Partiiny kontrol' deiatel'nosti administratsii. Politizdat, Moscow, 1983, 2nd edition.

PILIPENKO, N. N. *Dialektika neobkhodimosti i sluchainosti.* 'Mysl'', Moscow, 1981.

PISARENKO, S. M. *Tselevoe planirovanie i territorial'nye kompleksy.* 'Naukova dumka', Kiev, 1983.

Plenum Tsentral'nogo Komiteta KPSS 24–26 marta 1965 goda. Stenografichesky otchet. Politizdat, Moscow, 1965.

Plenum Tsentral'nogo Komiteta KPSS 14–15 iunia 1983 goda. Stenografichesky otchet. Politizdat, Moscow, 1983.

POKROVSKY, V. A. *Uskorenie nauchno-tekhnicheskogo progressa. Organizatsiia i metody.* 'Ekonomika', Moscow, 1983.

Politichesky dnevnik. 1964–70. Fond im. Gertsena, Amsterdam, 1972.

Politichesky dnevnik II. 1965–70. Fond im. Gertsena, Amsterdam, 1975.

POPOVSKY, M. *Delo akademika Vavilova.* 'Ermitazh', Ann Arbor, 1983.

POPOVSKY, M. *Science in Chains. The crisis of science and scientists in the Soviet Union today.* Collins and Harvill, London, 1980.

POPOVSKY, M. *Upravliaemaia nauka.* Overseas Publications Interchange, London, 1978.

PRAVDIN, A. (interviewed by M. Matthews). 'Inside the CPSU Central Committee', *Survey*, vol. 20, no. 4 (93), Autumn 1974, pp. 94–104.

Problemy rukovodstva nauchnym kollektivom. Opyt sotsial'no-psikhologicheskogo issledovaniia. 'Nauka', Moscow, 1982.

PUSHCHAROVSKY, Iu. M. and IANSHIN, A. L. *Tektonicheskoe razvitie zemnoi kory i razlomy.* 'Nauka', Moscow, 1979.

RABKIN, Y. M. 'On the origins of political control over the content of science in the Soviet Union' (an interview with A. Kol'man), *Canadian Slavonic Papers*, vol. 21, no. 2, 1979, pp. 225–37.

RABKIN, Y. M. 'The study of science', *Survey*, vol. 23, no. 1 (102), Winter 1977–78, pp. 134–45.

RAHR, A. 'Biographies of Politburo members', *Radio Liberty Supplement*, no. 1, 2 March 1984.

RAHR, A. 'Brezhnev's former party theorist Trapeznikov retires on pension', *Radio Liberty*, RL 350/83, 19 September 1983.

RAHR, A. 'Igor' Andropov to be new Soviet Ambassador to Greece', *Radio Liberty*, RL 281/84, 19 July 1984.

RAHR, A. 'New Central Committee departments formed', *Radio Liberty*, RL 229/83, 14 June 1983.

RAIKOVA, D. D. 'Osobennosti otnosheniia uchenykh k razlichnym sredstvam nauchnoi kommunikatsii', in A. A. Zvorykin (ed.), *Nauchny kollektiv. Opyt sotsiologicheskogo issledovaniia*. 'Nauka', Moscow, 1980, pp. 79–102.

Raionny komitet partii. Politizdat, Moscow, 1974, 2nd edition.

RASSUDOVSKY, V. A. *Gosudarstvennaia organizatsiia nauki v SSSR*. 'Iuridicheskaia literatura', Moscow, 1971.

REMNEK, R. B. (ed.). *Social Scientists and Policy Making in the USSR*. Praeger, New York, 1977.

REZNIK, S. *Doroga na eshafot*. 'Tret'ia volna', Paris and New York, 1983.

RIGBY, T. H. 'The CPSU from Stalin to Chernenko: membership and leadership', paper presented to a conference in honour of Lloyd Churchward, 'Forty years of Soviet studies', Melbourne, October 1984.

RIGBY, T. H. 'A conceptual approach to authority, power and policy in the Soviet Union', in T. H. Rigby, A. Brown and P. Reddaway (eds), *Authority, Power and Policy in the USSR. Essays dedicated to Leonard Schapiro*. Macmillan, London and Basingstoke, 1983, ch. 2.

RIGBY, T. H. 'Politics in the mono-organizational society', in A. C. Janos (ed.), *Authoritarian Politics in Communist Europe. Uniformity and diversity in one-party states*. Institute of International Studies, University of California, Berkeley, 1976, pp. 31–80.

RIGBY, T. H. ' "Totalitarianism" and change in economic systems', *Comparative Politics*, vol. 4, no. 3, April 1972, pp. 433–53.

RIGBY, T. H., BROWN, A. and REDDAWAY, P. (eds). *Authority, Power and Policy in the USSR. Essays dedicated to Leonard Schapiro*. Macmillan, London and Basingstoke, 1983.

RIGBY, T. H. and MILLER, R. F. *Political and Administrative Aspects of the Scientific and Technical Revolution in the USSR*, Occasional paper no. 11, Department of Political Science, RSSS, Australian National University, Canberra, 1976.

ROCCA, G. L. ' "A second party in our midst": the history of the Soviet Scientific Forecasting Association', *Social Studies of Science*, vol. 11, no. 2, May 1981, pp. 199–247.

RUBLE, B. A. 'Romanov's Leningrad', *Problems of Communism*, vol. 32, no. 6, November–December 1983, pp. 36–48.

SAKHAROV, A. D. *Sakharov Speaks*. Collins and Harvill, London, 1974.

SAKHAROV, S. V. *Primeniaia individual'ny podkhod*. Politizdat, Moscow, 1982.

SALOV, V. I. 'Iz istorii Akademii nauk SSSR v pervye gody Velikoi Otechestvennoi voiny (1941–1943)', *Istoricheskie zapiski*, Izd-vo AN SSSR, Moscow, 1957, pp. 3–20.

SCHAPIRO, L. *The Communist Party of the Soviet Union*. Eyre and Spottiswoode, London, 1970, 2nd edition.

SCHMITTER, P. 'Still the century of corporatism?', *Review of Politics*, vol. 36, no. 1, January 1974.

SCHWARTZ, J. J. and KEECH, W. R. 'Group influence and the policy process in the Soviet Union', *American Political Science Review*, vol. 62, no. 3, September 1968, pp. 840–51.

SHARENKOV, S. V. *Raikom i nauchno-tekhnicheskaia intelligentsiia*. Lenizdat, Leningrad, 1981.

SHCHERBAKOV, A. I. *Effektivnost' nauchnoi deiatel'nosti v SSSR. Metodologichesky aspekt*. 'Ekonomika', Moscow, 1982.

SHCHERBITSKY, B. V. and V. S. Tarasovich. *Planovoe upravlenie razvitiem nauki i tekhniki v soiuznoi respublike*. 'Naukova dumka', Kiev, 1981.

SHRAGIN, B. 'Oppozitsionnye nastroeniia v nauchnykh gorodakh', *SSSR. Vnutrennye protivorechiia*, no. 1, 1981, pp. 100–20.

SHTROMAS, A. *Political Change and Social Development. The case of the Soviet Union*. Lang, Frankfurt/Main and Bern, 1981.

SHUBKIN, V. N. 'Sotsiologicheskie problemy vybora professii', in G. V. Osipov and J. Szczepanski (eds), *Sotsial'nye problemy truda i proizvodstvo*. 'Mysl'', and 'Ksiazka i wiedze', Moscow and Warsaw, 1966, pp. 23–38.

SIGOV, I. I. *Kompleksnaia programma nauchno-tekhnicheskogo progressa (Metodologicheskie osnovy razrabotki)*. 'Nauka', Leningrad, 1983.

SIMES, D. K. 'National security under Andropov', *Problems of Communism*, vol. 32, no. 1, January–February 1983, pp. 32–9.

SISNEV, V. A. and STEPANOV, T. T. (eds). *Narodny kontrol' v sfere nauki, obrazovaniia, kul'tury, zdravookhraneniia*. 'Iuridizdat', Moscow, 1980.

SKILLING, H. G. 'Soviet and communist politics: a comparative approach', *Journal of Politics*, vol. 22, no. 2, May 1960, pp. 300–13.

SKILLING, H. G. and GRIFFITHS, F. (eds). *Interest Groups in Soviet Politics*. Princeton UP, New Jersey, 1971.

SLIUSARENKO, A. I. *KPSS i progress nauki v razvitom sotsialisticheskom obshchestve*. 'Vishcha shkola', Kiev, 1978.

SMITH, G. B., MAGGS, P. B. and GINSBURGS, G. (eds). *Soviet and East European Law and the Scientific-Technical Revolution*. Pergamon, New York, 1981.

SMOLKINA, A. A. *Deiatel'nost' KPSS po vosstanovleniiu i razvitiia nauchno-tekhnicheskogo potentsiala Leningrada (1945–1965gg)*. Izd-vo Leningradskogo un-ta, Leningrad, 1983.

SOBOLEV, G. L. *Uchenye Leningrada v gody Velikoi Otechestvennoi voiny, 1941–1945*. 'Nauka', Moscow and Leningrad, 1966.

SOLOMON, P. H. *Soviet Criminologists and Criminal Policy. Specialists in policy-making*. Macmillan, London and Basingstoke, 1978.

SOLOMON, P. H. 'Soviet policy making in comparative perspective', *Soviet Union/Union Soviétique*, vol. 8, pt 1, 1981, pp. 110–18.

SOLOMON, S. G. (ed.). *Pluralism in the Soviet Union. Essays in honour of H. Gordon Skilling*. Macmillan, London and Basingstoke, 1983.

SOLZHENITSYN, A. *Bodal'sia telenok s dubom. Ocherki literaturnoi zhizni*. YMCA, Paris, 1975.

Sovetskaia sotsiologiia. 'Nauka', Moscow, 1982, 2 vols.

VUCINICH, A. *Science in Russian Culture, 1861-1917* Stanford UP, 1970.

VUCINICH, A. *The Soviet Academy of Science*, Hoover Institute Studies, series E: Institutions, no. 3, Stanford UP, January 1956.

VUCINICH, A. *Social Thought in Tsarist Russia. The quest for a general science of society 1861-1917* University of Chicago Press, Chicago and London, 1976.

WISHNEVSKY, I. 'Soviet legislation and anonymous denunciations', *Radio Liberty*, RL 96/84, 2 March 1984.

WOOD, R. C. 'Scientists and politics: the rise of an apolitical elite', in R. G. Gilpin and C. Wright (eds), *Scientists and National Policy Making*, Columbia UP, New York and London, 1964, pp. 41-72.

WRINCH, P. N. 'Science and politics in the USSR: the genetics debate', *World Politics*, vol. 3, no. 4, July 1951, pp. 486-519.

YANOWITCH, M. and DODGE, N.T. 'The social evaluation of occupations in the Soviet Union', *Slavic Review*, vol. 28, no. 4, December 1969, pp. 619-41.

YORK, H. F. *The Advisor, Oppenheimer, Teller and the Superbomb*, Freeman, San Francisco, 1976.

YURENEN, S. 'Journal politics from Andropov to Chernenko', *Radio Liberty*, RL 299/84, 7 August 1984.

ZASLAVSKY, V. *The Neo-Stalinist State. Class, ethnicity and consensus in Soviet society*, Sharpe, New York, 1982.

ZEMTSOV, I. IKSI. *The Moscow Institute of Applied Social Research. A note on the development of sociology in the USSR*, Soviet Institutions Series, No. 6, The Hebrew University of Jerusalem, Jerusalem, April 1976.

ZIRKLE, C. *Death of a Science in Russia*, University of Philadelphia Press, 1949.

TSVETKOV, V.V. (ed.) *Sovershenstvovanie apparata gosudarstvennogo upravlenia. Konstitutsionnyi aspekt*, 'Naukova dumka', Kiev, 1982.

TUOMINEN, A. *The Bells of the Kremlin. An experience in communism*, University Press of New England, Hanover and London, 1983.

TURCHIN, V. *The Inertia of Fear and the Scientific World View*, Martin Robertson, Oxford, 1981.

V. I. Lenin, KPSS o razvitii nauki, Politizdat, Moscow, 1981.

VIDMER, R. F. 'Soviet studies of organization and management: a review of competing views', *Slavic Review*, vol. 40, no. 3, Fall 1981, pp. 404-22.

VLADIMIROV, L. *Sovetsky kosmichesky blef*, Posev, Frankfurt.

VODZINSKAIA, V. V. 'O sotsial'noi obuslovlennosti' in G. V. Osipov and J. Szczepanski (eds), *Sotsial'nye problemy truda i proizvodstve*, 'Mysl'' and 'Ksiazka i wiedza', Moscow and Warsaw, 1969, pp. 39-61.

VORONITSYN, S. 'Academicians clash over genetics', *Radio Liberty*, RL 284/81, 20 July 1981.

VORONITSYN, S. 'Attempts to reach a compromise on diversion of rivers', *Radio Liberty*, RL 313/84, 30 August 1984.

VOSLENSKY, M. *Nomenklatura. Anatomy of the Soviet ruling class*, Bodley Head, London, 1984.

VUCINICH, A. *Empire of Knowledge. The Academy of Sciences of the USSR (1917-1970)*, University of California Press, Berkeley and Los Angeles, 1984.

Index